CRYSTAL MODEL

Image courtesy of Warders Medical Centre Tonbridge, successors to John Gorham's practice. Thanks to Mrs C Burgess, Practice Manager for her assistance.

A SYSTEM FOR THE CONSTRUCTION OF CRYSTAL MODELS ON THE TYPE OF AN ORDINARY PLAIT

Dr. John Gorham, MCRS Eng, etc

Originally published in 1888

Introductory material by John Sharp

Tarquin Reprints

© Tarquin Publications 2007

ISBN 1 89961 868 6

Tarquin Publications
99 Hatfield Road
St Albans
AL1 4JP
www.tarquinbooks.com

Typeset by DPSL

Printed in Great Britain by Lighting Source

Also available as an e-book with multiple user licences – see website for details.

CONTENTS

ACKNOWLEDGEMENTS

Thanks for the following permissions to reprint articles and photographs:

The Pitt Rivers Museum, Oxford and Liz Yardley for the photograph of balls from the museum case on ball games, in the Introduction.

Joanna Gilmour and Richard Ahrens for photographs of their work in the Introduction.

Dr John Ford of Tonbridge and the staff of the Wellcome Trust Library, London, for biographical information

The Mathematical Association for allowing reprints of the two articles by A R Pargeter and James Brunton

Professor Jean Pedersen of Santa Cruz University for help with the bibliography

INTRODUCTION

Plaiting and braiding have a long history. They encompass decorating hair, weaving and basket-making as well as making knots. So it is surprising that John Gorham seems to have been the first person to have looked at the subject systematically. The main part of this book is a reprint of his work on plaiting crystal models has been an inspiration to many others working in this field, notably Robert Pargeter and Jean Pedersen. The Mathematical Association have kindly allowed us to reprint Robert Pargeter's article in the Mathematical Gazette together with James Brunton's follow up one. These extended Gorham's work and the ideas were extensively covered in *Mathematical Models* by Cundy and Rollett. It is through this book and Pargeter's article that Gorham's techniques have been kept alive. The book is rare and most references to it are only as a result of Pargeter having quoted it as his source.

Gorham was an interesting Victorian polymath and his biography is described separately. Also included is one of his papers on *The Kaleidoscopic Colour-Top* to show not only the breadth of his interests but also the aesthetic perspective of his work which is very much in evidence in the crystal models.

The earliest polyhedral nets are usually ascribed to Albrecht Dürer. However, he was just interested in cutting along the edges and so if you wanted to construct the model, you need to include tabs which are then glued. The beauty of Gorham's system is that no glue is needed. Dürer's net for a dodecahedron is shown in figure 1. It comes from his book *Unterweysung der Messung* (usually known as *The Painter's Manual* in English) which was an extensive geometry book published in various editions around 1500.

Dürer's book influenced many that followed, particularly books on perspective which used polyhedra extensively as drawing exercises. In Daniel Barbaro's *La Pratica della perpettiva* published in Venice in 1569, I thought I had found nets which were for plaiting like Gorham's ones. They look similar, but are in fact Dürer type nets for creating polyhedra by adding pyramids to the faces of other polyhedra. The example shown in figure 2 is the net for performing this operation on a tetrahedron to create a stella-octangula. The numbers define which of the triangles form each of the four pyramids.

Gorham's nets resemble Barbaro's but they are totally different. Gorham's nets are specially cut but this is not the only way to plait polyhedra. There are many cultures throughout the world who have woven polyhedra, for example as balls, but these generally do not have the edges of the polyhedra. These, however, are unlike Gorham's plaits because they are a series of straight strips because they are normally braided from a natural source such as a palm leaf. They usually result in a spherical object which has polyhedral symmetry and are often used as balls for playing games, a precursor of the soccer ball design. A set of such balls from the Pitt Rivers Museum in Oxford is shown in figure 3.

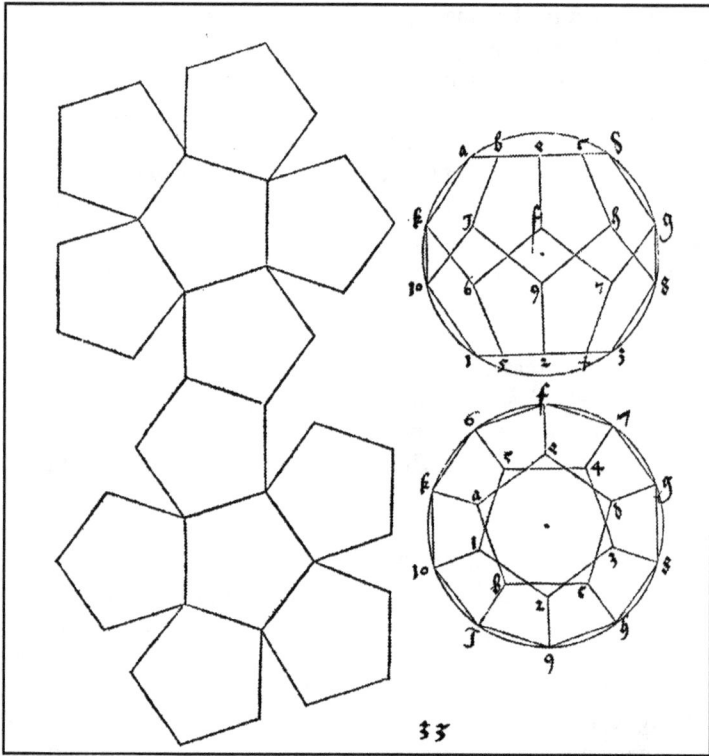

Figure 1.   Dürer's dodecahedron net

This has resulted in the progression to weaving rather than braiding a whole series of polyhedra from paper tape. This involves creasing the tape. For example, the cube can be made from three strips of five squares (one square on each strip overlaps) as in figure 4. There are many papers in the bibliographic references that use this technique.

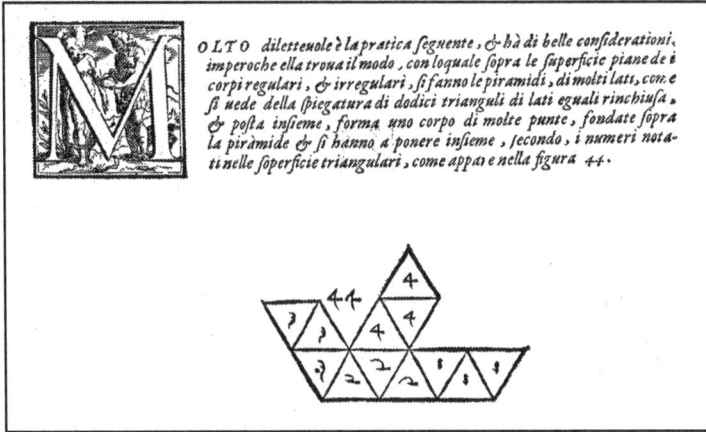

OLTO dilettenole è la pratica seguente, & hà di belle confiderationi, imperoche ella troua il modo, con loquale sopra le superficie piane de i corpi regulari, & irregulari, si fanno le piramidi, di molti lati, come si uede della spiegatura di dodici trianguli di lati eguali rinchiusa, & posta insieme, forma uno corpo di molte punte, fondate sopra la piràmide & si hanno a ponere insieme, secondo, i numeri notati nelle superficie triangulari, come appare nella figura 44.

Figure 2.   Barbaro's net

Figure 3.   The ball games case in the Pitt Rivers Museum Oxford.

You will observe that neither Gorham, nor Pargeter in his enhancement of the technique make the pentagonal dodecahedron or polyhedra with faces containing hexagons (like the truncated octahedron) or octahedrons (like the truncated cube). Such polygons are obtainable from continuous strips by making knots in them. Jiménez et al also make creases in the strips which mimic the creases in the knots. James Brunton followed on Pargeter's article with one on plaiting pentagonal dodecahedron which is also reprinted in this book.

It is still possible to obtain a cube with minimal folding by plaiting paper strips. The examples in figure 5, by Joanna Gilmour, also show how the material can have a marked effect on the artistic result.

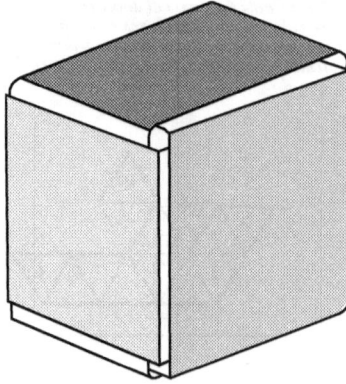

Figure 4.   Three strips forming a cube

Figure 5.   Plaited cubes by Joanna Gilmour

Further examples are found in the work of Richard Ahrens who uses plastic strapping used to wrap packages. Figure 6 shows a tetrahedron made this way.

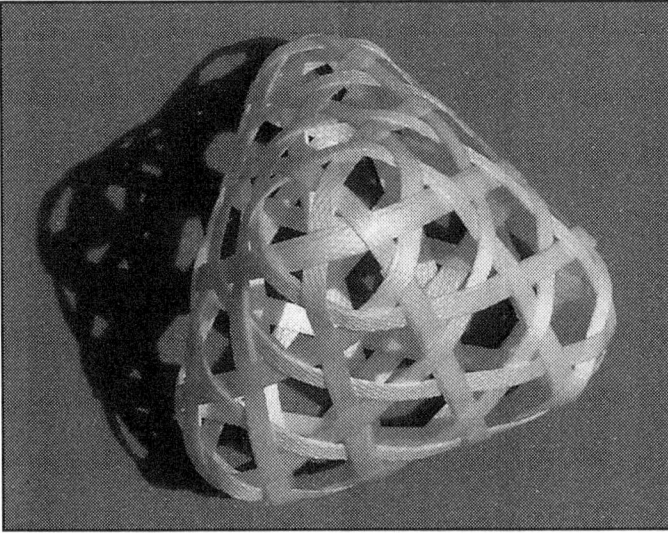

Figure 6.   Tetrahedron by Richard Ahrens

Figure 7 shows a further development where the instead of the polyhedron fitting on a sphere, it fits on a double torus.

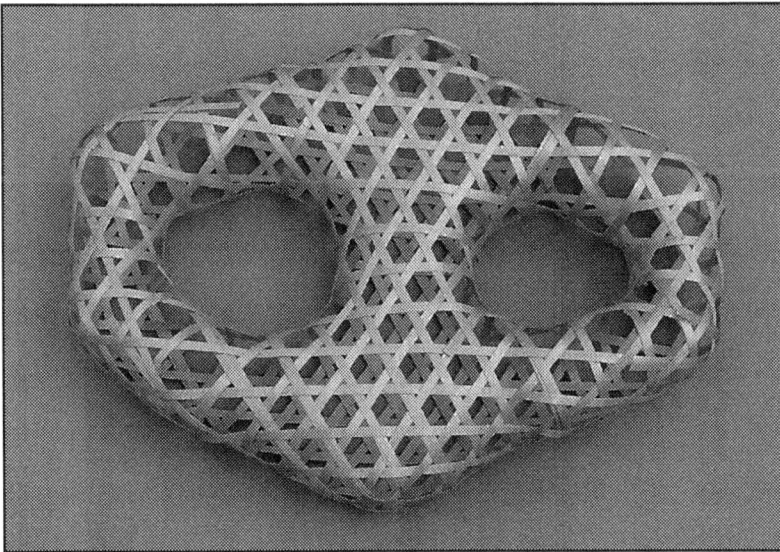

Figure 7.   Woven polyhedron based on a double holed torus

John Gorham's idea, as he describes in his preface, was to create an inexpensive set of models for use in teaching crystallography. Over a century later, as you can see by the large bibliography, the subject could be the basis of yet another book. One wonders what John Gorham would have made of the ideas he spawned.

DR JOHN GORHAM – A BIOGRAPHY

John Gorham was a distinguished doctor and polymath who was born in 1814 near to Tonbridge in Kent. He was educated at Tonbridge school and then became an apprentice to a local doctor William James West who came to Tonbridge from Saint Albans in 1815. West was the first in the uninterrupted line of doctors that we can trace to what is now Warders Medical Centre in the town. We are grateful for much of the following history to a present member of the centre Dr. John Ford.

In 1832 Gorham went to Guy's hospital to complete his medical training. After qualifying with the MRCS LSA in 1835, he stayed at Guy's and worked as a practitioner in Southwark, for further experience. This included taking Thomas Addison's outpatient clinics when Addison was away. Addison is famous for giving his name to the condition of underactivity of the adrenal glands which he described in 1855. Gorham gave several papers to Guy's Medical and Physical Society, and in 1838 published the first of many papers on medical and other subjects. The subjects included, Intussusception in Children (1838), A Case of Fungoid Disease of the Kidney (1838), and a case of Extraordinary Development of the Mammae in a Male (1839). Gorham joined the controversy in the medical journals about the propriety of removing ovarian cysts with Observations on the Propriety of Extirpating the Cyst in Some Cases of Ovarian Dropsy (1839).

By 1841 he was back in Tonbridge, in partnership with West. They both became deeply involved in town matters as well as medicine. This was very important to the practice for getting to know local worthies who hopefully might become patients. West's death from dropsy in 1848 was a tragedy for the town. Gorham took over the practice, which he ran single-handed with the help of numerous assistants until 1873. We know as little of the day-to-day work of Gorham as we do of other practitioners, for such work leaves little documentation.

In 1841, Gorham was appointed Medical Attendant to the Poor Law Union. It would have been difficult not to appoint him, as he had glowing testimonials from thirteen distinguished doctors from Guy's, including Richard Bright, Thomas Addison, Aston Key, and John Hilton. Whilst at Guy's a microscopical department was founded and it could have been this that fired Gorham's interest. He joined the Royal Microscopical Society and contributed several publications to its journal. These included papers on compound eyes in insects, venation in the leaves of umbelliferae and on the composite structure of simple leaves. He was a practical man, as in other papers he suggested ways of making casts of insect eyes and other objects in collodion for microscopic study and improving ways of preparing microscopic slides. His botanical interests were furthered by membership of The Royal Botanical Society.

By 1869 his interest had turned to teeth. He wrote a booklet on "The proper method of extracting teeth." It was full of common sense and advice, and pointed out that if the doctor learnt early on his career how to do it, "the practitioner might vie with the professional dentist in this department of surgery." This would have added to the doctor's income considerably and it is not surprising that the booklet ran to five editions.

His interest in teeth lead him to study some thousands of them supplied by a local dentist and Charles Fox at the London Dental Hospital. He published his findings in *The Medical Times and Gazette* and they were reprinted in *The British Journal of Dental Science*. Each was carefully measured and the average weight of each tooth calculated. All the lower teeth were lighter than their complementary upper ones. By measuring the distance from each tooth to the mid point of the jaw, he worked out the 'lever power' for each tooth. After many calculations he came to the conclusion that if the lower teeth were as heavy as the upper, an extra 200 grains of work would be necessary at each bite. The inescapable conclusion was that 'The Divine Artificer has designed the body perfectly and that the arrangement of the teeth was 'the best and wisest that could be devised'. This fundamentalist view accorded well with his Free Church views. He never really got involved with the Parish Church.

Gorham kept a very high profile in the town. He gave penny readings at the Mechanics Institute, which he had helped to found and played his flute in trios with his wife and daughter in concerts at The Public Hall. He became President of The Aesthetic Society and later, The Choral Society. He had many interests besides medicine. The British Medical Journal obituary says that "… science was the real bent of Mr. Gorham's clear and well balanced mind, and it is understood that a Fellowship of the Royal Society was long open to him had he cared to accept it. Crystallography was one of his special hobbies; and the science of colour equally interested him. He was thus often brought into contact with the late Sir William Herschell and other leaders of science and was elected a Fellow of the Physical Society of Guy's and of he Royal Botanical Society, and a member of the Mineralogical Society." He gave courses of lectures in the Public Hall in Tonbridge on mineralogy, colour and conchology. That on colour was 'not very well attended, but the company was select'.

It was the interest in mineralogy which gave rise to the present book. He read a paper on *A System for constructing Crystal Forms by the Plaiting of their*

Zones a few year after the book was published to the Mineralogical Society, the abstract of which is reprinted in the following pages.

Also included is a paper from Volume 1 of *Recreative Science: a Record and Remembrancer of Intellectual Observation* published in 1860 by Groombridge and Sons of Paternoster Row. This was patented (no 806, April 14 1858) as "Improvements in optical instruments by the revolution of which various designs or patterns may be produced by the eye." The abstract appears as "Tops: In chromatic tops, a blackened disc, with a design cut out of it, and a central hole larger than the top spindle, is placed above the usual coloured discs. To retard the disc with the design as the top spins, a piece of string may be fixed to its periphery." The device was manufactured and one was exhibited at the *Illusions* exhibition at the Hayward Gallery in London in 2004–5. It is a variation on the colour top of James Clerk Maxwell which Maxwell presented in a paper entitled, *Experiments on Colour, as perceived by the Eye, with remarks on Colour-blindness* to the Royal Society of Edinburgh in 1855 in his paper. His top demonstrated that any natural colour could be produced from the three primary colours-red, green and blue. Gorham must have been well read to have picked this up, but his inventiveness shows how he developed it.

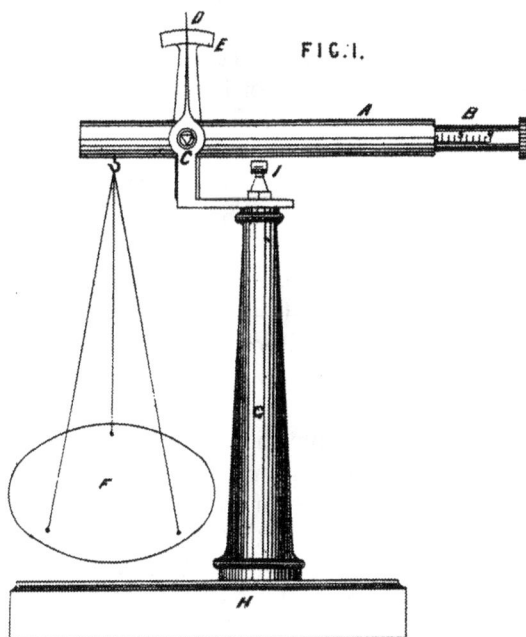

FIC.1.

There is another patent (no 3064 of 1880) which is a tubular or telescopic balance. A pan is suspended at the end of a tubular bar which has a scale sliding in it. The weight in the pan is determined by moving the scale out of the bar until a balance is achieved. So with this wide range of interests, it is no surprise that his son, John Marshall Gorham born in 1853, qualified and practised in London as an engineer. He attended Tonbridge School 1865–70 and then became an electrical engineer trading as Drake and Gorham of Victoria Ste London. He died at Hawkhurst, Kent in 1929.

In 1894 he retired and died of influenza in 1899, by which time the young man who had worked with Addison and Bright in the advances of the early nineteenth century was described as 'a doctor of the old school'.

BMJ OBITUARY

One of the oldest and most respected of the members of the profession, Mr. JOHN GORHAM, has fallen a victim to influenza, and after a week's illness passed peacefully away on December 13th at Tonbridge, Kent. He was born in 1814 near that town, and had consequently completed his 85th year. He was educated at Tonbridge School, and received his medical training at Guy's Hospital. He passed "the Hall" in 1835, and became M.R.C.S. in 1839. He then settled at Tonbridge, and joined Dr. West, a surgeon of whom it is reported that he performed ovariotomy several times before it was even attempted in London. Mr. Gorham contributed many papers to the medical journals, all characterised by a pure classical style of writing. Amongst these were papers on Intussusception in infants, Ovarian Dropsy, Respiration and Pulse in the Infant, and the Leverage of tile Lower Jaw. He was also the author of a Manual on the Proper Mode of Extracting Teeth, the third edition of which appeared in 1889. But science was the real bent of Mr. Gorham's clear and well-balanced mind, and it is understood that a Fellowship of the Royal Society was long open to him had he cared to accept it. Crystallography was one of his special hobbies; and the science of colour equally interested him. He was thus often brought into contact with the late Sir William Herschell and other leaders of science, and was elected a Fellow of the Physical Society of Guy's and of the Royal Botanical Society, and a Member of the Mineralogical society. Amongst other papers between 1854 and 1889 he wrote the following: Unfrequented Paths in Optics, Part I; Light from a Pinhole, Part II; Light from Fissure; A System for the Construction of Crystal Models of the Type of an Ordinary Plait; The Rotation of Coloured Discs; Mode of Calculating Hexagonal Facets in Cornea of the Eye in Insects; The Diascope, a New Optical Instrument; Rudiments of Colour by Rotation; The Stereograph, for Delineating Solids in Space; Venation in Leaves of Umbelliferae; On the Composite Structure of Simple Leaves; The Tubular Balance; On the Blending of Colours by the Sole Agency of the Sensorium; The Pupil Photometer; etc. He was also the inventor of the entoscope, an instrument for measuring the size of our own pupil from within.

A SYSTEM FOR CONSTRUCTING CRYSTAL FORMS BY THE PLAITING OF THEIR ZONES

By J. GORHAM, M.R.C.S.

Reprinted from the Mineralogical Magazine 1892, Volume 9 pp235–6.
[Read January 27th, 1891.]

In this paper the author described his method of constructing crystal models of paper or cardboard by means of plaiting. In the ordinary crystal "nets" (such as, according to Mr. Gorham, were constructed by himself and shown at the Royal Society as early as 1847) the faces of the model are drawn in juxtaposition on cardboard and cut out, some of the lines being only partially cut through; the net is then folded up so as to form a solid figure, and the faces are united at their edges by gum.

In the models described and exhibited by Mr. Gorham, the faces are so drawn as to be disposed along strips; these strips are united to each other at one end, and can then be folded together by a process resembling plaiting, in which the strips pass alternately under and over each other. In this way models are made precisely like those constructed from the ordinary nets, but requiring no cement; models which can, moreover, be taken to pieces (unplaited) and laid flat, or re-constructed whenever desired.

Each strip contains the faces lying in one zone, and the number of strips employed by Mr. Gorham is either three or four; the cube, for instance, is formed by the plaiting together of three, and the dodecahedron by the plaiting together of four strips.

Their order of succession is the same for each model whatever its form. Thus the strips of square faces from which the cube is constructed are three in number, and if they be denoted by the numerals 1, 2, 3 respectively, the process of plaiting always pursues the following order:-

    1 passes over 2
  then 3 passes over 1
  then 2 passes over 3 and so on.

Any model consisting of three strips is plaited in the same order.

A peculiar (and a new) feature of the models exhibited ... is that their faces
are traversed by lines; one strip is marked by a simple line, the second by a
double, and the third by a triple line. These lines fulfil a twofold purpose:
(1) by their direction they show the number and disposition of the zones in
each crystal (and indicate the order in which the strips are to be plaited); and
(2) the alternate appearance and disappearance of the lines, that is to say
their appearance on alternate faces of the model, is strongly suggestive of
their being intertwined like a plait, as is of course the fact. It is to this alone
that that the model is indebted for its solid form and its power of retaining
the same.

THE KALEIDOSCOPIC COLOUR-TOP

By JOHN GORHAM, ESQ., M.R.C.S., OF TUNBRIDGE

Pages 89–93 from Volume 1 of *Recreative Science: a Record and remembrance of Intellectual Observation* published in 1860 by Groombridge and Sons of Paternoster Row.

Have you not seen in the experiments with Armstrong's electric machine, at the Polytechnic Institution, a series of sparks passed along a line of insulated conductors in close proximity, which in their rapidity of passage appeared like an unbroken stream of light? The lecturer has explained that the fluid passes in distinct and separate flashes, and, by a manipulation for the purpose, has proved that such is the case. Yet, when his regiment of brass knobs is properly arranged, the sparks pass from one to the other in such rapid succession that you cannot distinguish one from another, and you see a line of light with apparently not a break in it. This is sufficient to prove that we must not depend wholly on the impressions conveyed to us by our senses, and it also proves that apparitions of objects remain on the retina of the eye after the act of real vision is at an end. Go to a factory and choose for experiment a rapidly rotating fly-wheel. When the machinery is at rest, paint the rim of the wheel with zigzag lines of blue and white, side by side, all round, very distinctly, so that there can be no question as to there being two colours unmixed, though in close proximity.

As soon as the machinery is again set in motion, look at that fly-wheel. The zigzags have vanished; there is not a line of blue or white about it, but it is painted of a uniform gray colour! No; it is as you left it, in zigzags of two colours, but by its rapid action the eye fails to detect them, and sees the two as one, and the only point of interest about it that we now mean to mention is, that the change has taken place in the organs of vision, not in the painting of the wheel.

Chevreul has analyzed the laws of colour with a view to point out how harmonies and contrasts are best produced, and in effecting this object he has taken occasion to explain how colours reciprocally act on each other, so as to heighten, diminish, or otherwise alter the proper effects of each when variously brought into juxtaposition. In this case, too, the changes take place in the sense of vision, not in the colours themselves. But Chevreul's experiments with strips of coloured paper, instructive as they are do not induct us very far into the mystecious region colours and harmonized tints, and we must have recourse to motion as an agency powerfully operative in effecting such modifications.

The Kaleidoscopic Colour-top, is intended for the express purpose of effecting changes in colour by motion, and promises to become most instructive to the artist and the philosopher, as well as a source of high gratification as a toy for the family circle.

What is the colour-top? Strictly it is but a coloured card made to revolve horizontally on a vertical axis; or rather, the card is an addendum to the simple apparatus whose business it is to spin when desired, and the card plays the part of the fly-wheel we just now instanced as an example of distinct colours blended into one, different from either of its components, by rapid rotation.

This is a mere elementary hint of what the colour-top is capable of in regard to blending of hues and colours. In the hands of the man of science it is an instrument of wonderful power, and, like a magician's shuttle, weaves as it were the rainbow itself into patterns of endless variety, in which form as well as colour plays its part in the production of extraordinary appearances.

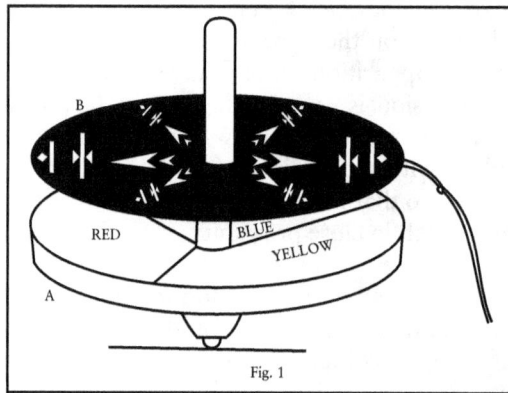

Fig. 1

In a description published in the pages of the "Microscopical Journal" of January last, the following particulars are given of the simplest use of the instrument:-

Let A (Fig. 1) be a disc having the three primary colours, yellow, red, and blue, evenly distributed in sectors of 120 on its surface, and let this disc be fixed to the colour-top and rotated. Let B be a second disc, blackened on its upper surface, and having a central aperture somewhat larger than the spindle of the top, as well as a pattern of six or eight rays cut completely through and out of its substance, so that when held vertically over the disc A it will admit of the colours being seen through such pattern when viewed from above. If now the wheel is set in motion, and the disc B allowed to drop down upon the spindle of the colour-top, it will be held in contact with the spindle by centrifugal force, and will revolve in a plane parallel with the wheel, in the same direction and with the same velocity. In this case the colours on A appear mixed, while the pattern on B is effaced. But if the motion of B is retarded, and at the same time broken up into a series of rapid and regular jerks, or more properly isochronous vibrations, while the eye is held vertically over the spindle, each pattern is retained for an instant before the eye, yet sufficiently long to form a distinct image; and owing to the rapidity of the vibrations, a whole circle of images is thus portrayed on the retina of the eye before the image from the first vibration is effaced. In a like manner, the colours of the disc A, which are perceived only as they are transmitted through the open designs in the disc B, appear in their primitive purity or unmixed state, the colour in one sector being reflected through a given ray of the pattern before the arrival of the colour from that sector by which it is immediately succeeded. Hence, both pattern and colours appear multiplied, thus producing the combinations.

From the construction of the instrument; about six revolutions of the disc A occur to one single revolution of the disc B. When these relative velocities are maintained, five groups of all the colours distributed on the colour-disc are seen occurring in the order of their arrangement on the disc, and repeated in perfect symmetry in the various openings of the patterns. In this way the most beautiful variations may be effected by using different colours in various proportions on the disc A; for, however numerous the colours, each colour is reflected through its proper opening in the disc B, at a given interval of space and time, without the slightest irregularity or confusion. Fig. 2 represents a non-rotating disc, and Fig. 1 the same during rotation.

In order to retard the motion of the disc B, and, at the same time, to produce the vibrations, the central aperture is made sufficiently large to admit of free motion on the spindle; and there is appended, at or near its circumference, a light weight, such as a piece of string or silk, which, by its impulses on the atmosphere during rotation, both retards the motion and produces the vibrations. The proportion of the diameter of the central aperture of the disc to that of the spindle of the colour-top is about as four to three-an aperture of four-tenths of an inch, for example, to a spindle three-tenths of an inch, is very effective.

As in the kaleidoscope, so in this instrument, there is only one position for the situation of the eye with respect to the discs where the symmetry of the combinations is perfect, namely, vertically over the spindle of the top, so that the whole of the circular field can be distinctly seen.

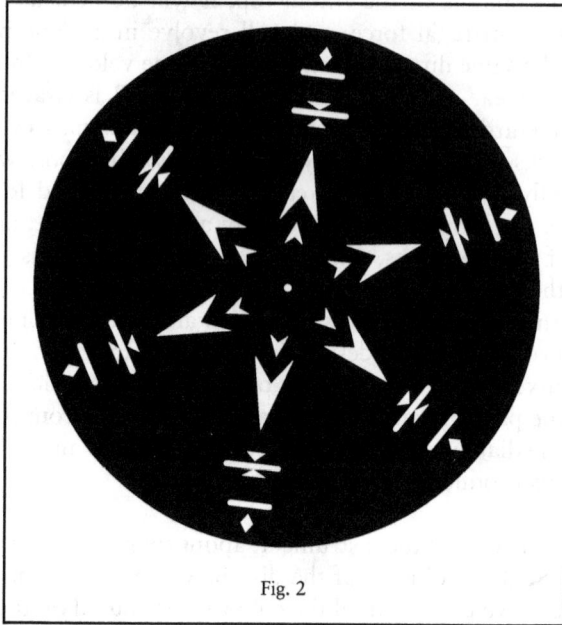

Fig. 2

For the purpose of giving variety to the figures formed by the instrument, an assortment of fantastic patterns on separate discs may be constructed to take the place of the disc B; the disc A is also furnished with different colours. A disc of pure white forms exquisite gray combinations, with every conceivable variety of light and shade; a disc of white and green, in equal portions, is resolved into these elements, and their multiplication into a composite form is very agreeable to the eye. An elegant arrangement, whereby patterns may be coloured in the most attractive manner, is composed of half a disc of blue, and the rest of white, green, and red, in equal proportions.

To the artist this instrument will prove of service in enabling him to select any number of colours in any proportion, and to adjust them instantaneously into a symmetrical pattern. He will thus be enabled to ascertain whether his selection of colours is harmonious, and he will be able to put to the test some of those beautiful harmonies which are introduced into the more recent works on colour. The following arrangement, adapted from Sir J. Gardner Wilkinson's work on Colour and Taste, will serve as an example:-

On the disc A arrange, in the following order and proportions, the colours-Blue 129; Scarlet, 42; Crimson, 37; Orange, 19; Yellow, 52; Green, 41; White, 40.

Place such a disc on the colour-top, and rotate; during rotation drop down upon the spindle any of the black perforated discs with strings appended to them: the colours of the disc A will now reappear multiplied, and presenting themselves here and there, filling up the open patterns of the perforated disc B, and forming a *polychrome* ornament, the beauty of which can scarcely be surpassed.

The pictures thus presented to the eye are very beautiful. Their charm would appear to depend partly upon their being reflected to the eye through a perfectly black medium, which imparts brilliancy and illumination to the colours, and partly upon their being exhibited in a state of motion, the *apparent* and *real* direction of which bear no relation to one another. While the disc is actually performing one hundred revolutions per minute, and vibrating about thirty times during each revolution, the combinations themselves often *appear* hanging in space, trembling without progressing, or perfectly motionless, or gliding round in a direction contrary to that of their real motion. These frequently recurring and illusory changes excite the curiosity, give animation to the pictures, and confer an ethereal brightness, vivacity, and splendour upon them, which is altogether and peculiarly their own.

In the preparation of the discs intense washes of these pigments are chosen, and as few are required to produce an almost incalculable number of changes, the rationale of the colour-top is as simple as that of the kaleidoscope; and, in the production of colours by it, we are enabled to refer such colours to their constituents, and thus have a key to the mixing of colours for every purpose required by art. It is strange, however, that though red and yellow produce orange by rotation, and full tones and half tones result in others intermediate between them, yellow and blue, however proportioned to each other, have never yet produced even a tolerable green. This fact stands in the way at present of any attempt to form a nomenclature of colours which shall have any philosophical value. In all other respects the colour-top demonstrates the laws of contrast, evokes the complementaries, and enables the operator to blend colours in softer graduations than can ever be accomplished by the pencil. Indeed, for merely visual effects, and as giving an index to the action of colours, hues, tints, and shades, upon each other, it obviates the necessity of mixing pigments until it has determined previously the nature of the mixture required for any given tone in all the range of colouring, with the strange exception of the tones of green. Let us hope that this exception may some day be found capable of explanation; at present it is a mystery.

Messrs. Smith, Beck, and Beck, 6, Coleman Street, and Messrs. Elliott Brothers, 30, Strand, are the accredited agents for this curious and instructive instrument.

The Kaleidoscopic Colour-top is also mentioned in John Henry Pepper's *Boy's Playbook of Science*, London 1860.

"The invention by John Graham (sic), of Tunbridge, is designed to show that when white or coloured light is transmitted to the eye through small openings cut into patterns or devices, and when such openings are made to pass before the eye in rapid successive jerks, both form and colour are retained upon the nerve of the visual organ sufficiently long to produce a compound pattern, all parts of which appear simultaneously, although presented in succession."

BIBLIOGRAPHY OF GORHAM'S PAPERS

Obituaries of him can be found in the *Lancet* 1900 1 p140 and *The British Medical Journal* 1900 1 Jan 20.

There is an obituary of him in the *Tonbridge Free Press* December 16 1899, but it does not talk much about his interests.

He appears in the *Medical Directory* from 1850 until 1893 and the list of papers increases as he gets older.

John Gorham, Remarks on the cornea of insects, *Quarterly Journal of Microscopical Science* 1 1853, p76–84

John Gorham, Mode of calculating hexagonal facets in the cornea of the eye in insects, Quarterly *Journal of Microscopical Science* 2 1854

John Gorham, The Diascope, a new optical instrument, *Quarterly Journal of Microscopical Science* 2 1854

John Gorham, On magnifying power of short spaces, illustrated by the transmission of light through minute apertures. *Quarterly Journal of Microscopical Science* 2 1854, 218–234, 3 1855 p 1–15 and 4 1856 p27–44

John Gorham, The Rotation of Coloured Discs applied to facilitate the study of the Laws of Harmonious Colouring, and to the multiplication of Images of Objects into Kaleidoscopic Combinations. *Quarterly Journal of Microscopical Science* 1859, 7, p69–78

John Gorham, The Kaleidoscopic Colour Top, *Recreative Science: a Record and Remembrancer of Intellectual Observation* 1 1860, p 89–93, (contents reproduced in this book)

John Gorham, On a particular distribution of vein in leaves of the natural order umbelliferae, *Micros Soc Trans* XVI 1868, p 14–25

John Gorham, On the composite structure of simple leaves, *Monthly Microscopical Journal* 1 1869, p 155–69

John Gorham, Rudiments of colour by rotation, *Transactions Royal Microscopical Society* 1869?

John Gorham, The Stereograph for delineating solids in space, *Transactions Royal Microscopical Society* 1869?

John Gorham, On the blending of colours by the sole agency of the sensorium, *Brain* **4** 1882, 467–72

John Gorham, The pupil photometer *Proc Royal Society* 37, 1884 pp 425–6

John Gorham, A system of constructing crystal form by the plaiting of their zones, *Mineralogical Magazine* **9**, 1892 p 235–6 (contents reproduced in this book)

John Gorham, "Concerning the early days of ovariotomy", *Lancet* **103** (issue 2639) 28 March 1874, p 440–441

John Gorham, "Extraordinary development of mammae in the mail", *Lancet* **34** (issue 882) 25 July 1840, p 637–638

John Gorham, "Excision in ovarian dropsy", *Lancet* **33** (issue 852) 28 December 1839, p 506–7

John Gorham, "Observations on the property of extirpating the cyst in some cases of ovarian dropsy", Lancet **33** (issue 843) 26 October 1839, p 155–161

Apart from the crystals models reproduced here in facsimile, he also wrote a small book on Tooth extraction, which was published by H K Lewis and went through a number of editions from 1870 to 1893:

Tooth extractions: A manual on the proper mode of extracting teeth with a table in parallel columns of all the teeth, the instruments required for their extraction and the most approved methods of using them.

BIBLIOGRAPHY OF PAPERS ON PLAITING AND BRAIDING POLYHEDRA

Wendy Arbeit, "What are fronds for?"

James Brunton, "The plaited dodecahedron", *Mathematical Gazette*, Feb 1960, p12 (reprinted in this book)

James K. Brunton, "Polygonal knots", *Mathematical Gazette*, Dec 1961, p 299–302, (reprinted in this book)

Joanna Gilmour, "Plaiting a cube", *Infinity: Mathematical puzzles and diversions* (Tarquin Publications), 1, 2005 p 32–35

I Jiménez, G Pastor, A Ferriero, M Torres and E Gancedo, "The construction of Platonic Bodies from constant width continuous strips", *International Journal of Mathematical Education, Science and Technology*, 21 (1) 1990, p 213–218

I Jiménez, G Pastor, and M Torres, "The construction of stellated regular polyhedra from constant width strips", *International Journal of Mathematical Education, Science and Technology*, 21 (2) 1990, p 37–50

Edouard Lucas, "Le Noeud de Cravate" in *Récréations mathématiques*, vol 2 p 202–4

F V Morley, "A note on Knots", *American Mathematical Monthly*, 31, 1924, p 237–9

A. R. Pargeter, "Plaited polyhedra", *Mathematical Gazette*, May 1959, p 88–101 (reprinted in this book)

Jean J Pedersen, "Plaited platonic puzzles", *Two-Year College Mathematics Journal*, 1973, p 22–37

Jean J Pedersen, "Platonic solids from strips and clips", *Australian Mathematics Teacher* , 30 1974, p 130–33

Jean J Pedersen, "Braided rotating rings", *Mathematical Gazette*, **62** 1978, p 15–18 (reprinted in *Oregon Mathematics Teacher* March 1979 p 13–15)

Jean J Pedersen, "Visualising parallel divisions of space", *Mathematical Gazette*, **62** 1978, p 250–262

Jean J Pedersen, "Some isonemal fabrics on polyhedral surfaces", in *The Geometric Vein, the Coxeter Festschrift*, Springer Verlag 1981, p 99–120

Jean J Pedersen, "Geometry: The unity of theory and practice", *Mathematical Intelligencer* **6** 1984 p 54–56

Jean J Pedersen, "Why study polyhedra?" in *Shaping Space: A polyhedral approach* edited by Marjorie Senechal and George Fleck, Birkhauser, Design Science Collection, Boston 1988 p 133–147

Jean J Pedersen and Peter Hilton, "Constructing Jennifer's puzzle", *New Zealand Mathematics Teacher* **24**, 1987 (4) p 13–21

Jean J Pedersen, "Braiding tetrahedra and Cubes from US Bills", (letter to the editor) *Mathematics Magazine* **61** 1988 p 270

Jean J Pedersen and Peter Hilton, "Symmetry in Practice, Recreational Constructions", on the *Vismath* website at http://members.tripod/vismath

Jean J Pedersen, "Parallel Divisions of space", in *Mathematical Adventures for Students and Amateurs*, edited by David Hayes and Tatiana Shubin, MAA Spectrum Series 2004, p 113–117

Jean Pedersen, Derek Holton and Peter Hilton, *Mathematical Vistas – in a room with many windows*, Undergraduate texts in mathematics, Springer Verlag 2002

Rabe-Rüdiger von Randow, "Plaited Polyhedra", *Mathematical Intelligencer*, 26 2004 p54

PLAITED POLYHEDRA

A R PARGETER

Reprinted with kind permission of the Mathematical Gazette. Originally printed in May 1959 p 88–101.

1. In Cundy and Rollett's invaluable book on *Mathematical Models* [1], the authors begin their Chapter on Polyhedra by remarking that "The most suitable, and in many ways the most attractive, subject for an experiment in the construction of mathematical models is a set of polyhedra." Various methods are in general use to produce finished models of polyhedra for the showcase, and similar methods form the basis of those considered suitable for constructing the simpler polyhedra in the classroom. That most commonly adopted is to draw the net of the required solid on a sheet of cardboard, allow for tabs as necessary, cut it out, score the creases half-through, fold up, and stick the tabs with some suitable quick-drying cement. For classroom use the major drawback to model making on these lines is the time-consuming and potentially messy process of sticking; and it is the object of this article to develop a method which dispenses with the use of paste or cement altogether. In fact I shall show how, using paper and scissors only, firm, neat models can be made, which in the case of the simpler solids could easily be produced in the class-room, and which in the case of the more complex polyhedra reveal by their construction in a striking manner some of the geometrical relationships of these solids.

2. My attention was first drawn to this idea by a reference in *Multi-Sensory Aids in the Teaching of Mathematics* [2] to an out-of-print book by John Gorham, *Plaited Crystal Models* [3], published in 1888. The article in *Multi-Sensory Aids* is too condensed to be very illuminating, and is illustrated by a single and not very well chosen example, and I did not attempt to pursue the matter further until I managed to lay hands on a copy of Gorham's book, when my interest was fully aroused. John Gorham was a medical man with an interest in crystallography. He states in his Preface: "It is now some forty years since I had the honour of demonstrating before the Royal Society in London *A System for the Construction of Crystal Models Projected on Plane Surfaces*. These figures folded into the required form, and subsided into a level at pleasure—they were easily moulded into shape by bringing their edges into apposition with the fingers, and were as easily transferred from place to place when flattened in a portfolio ..." He naturally confines his attention to those forms in which a crystallographer would be interested,

namely the cube, regular tetrahedron and regular octahedron, numerous irregular versions of these (including what he terms rhombohedra), triangular and hexagonal prisms (the latter with and without pyramidal ends), and the rhombic dodecahedron. Not all of these solids are of equal interest to the mathematician, for whom there are obvious gaps, in particular the regular dodecahedron and icosahedron. I therefore set out to see whether Gorham's methods could be applied to these solids; and, succeeding, to extend them to the construction of Archimedean, dual, and stellated polyhedra. The time at my disposal has not permitted me to carry out the construction of more than a small selection of all the possible models, but in many cases it has been sufficient to observe that an effective construction *is* possible without being obliged to carry it out to verify the assertion: I shall show later how once a polyhedron has been constructed by a "compound plait", a simple modification at once leads to the construction of a range of related (though possibly quite different) polyhedra, and so it is only necessary to make one member of the set to have available the procedure for constructing the others. I shall show that it is theoretically possible to plait *any* polyhedron; but in some cases the method may be impracticable owing to too great complexity. Gorham does not appear to have made any attempt to generalize his methods, or to develop the underlying theory—in fact, as one extract quoted in *Multi-Sensory Aids* shows, his approach was primarily experimental; and he remained content—so far as one can judge from his book—with those models in which he (as a crystallographer) was interested. I shall show later how methods for plaiting any desired solid can be worked out systematically.

3. It is fitting to preface the descriptive part of this article with Gorham's own measured words: "It is a property inherent in the ordinary plait that its constituent parts shall cohere compactly without adventitious aid. This involves definite arrangement; and if this arrangement is studiously followed in the construction of models, their faces will in like manner become coherent, and, when plaited, will assume a solid form, and maintain it without extraneous assistance."

4. In proceeding to describe in detail the method of making plaited models, I shall use the term "net" to designate the net of a polyhedron as prepared for plaiting, and not the net as ordinarily used for a conventional model. It should also be noted that as there is some lack of uniformity in the naming of a few polyhedra—especially the Kepler-Poinsot star polyhedra—I shall adhere to those names adopted in *Mathematical Models*. It will in fact be assumed that the reader of this article is already acquainted with the names and forms of the Pythagorean, Kepler-Poinsot, Archimedean, dual, and stellated polyhedra: it would take too long to enumerate them here, and the information can be found, in varying degrees of completeness, in several well-known books (e.g. Rouse Ball's *Mathematical Recreations and Essays* [4], but especially *Mathematical Models* already referred to.)

5. In setting out to make the actual models the following points must be noted:

5.1. In the nets as here depicted, [see p 27 of this edition] *thick* lines should be cut, *thin* lines should be creased; a short stroke across a line indicates a reversed crease; *dotted* lines are sometimes inserted to indicate structure or to facilitate description.

5.2. On account of the amount of overlapping involved, *firm paper (e.g. cartridge paper) is preferable to cardboard* (although very thin cardboard could be used for large models).

5.3. Assuming that the net has been accurately drawn and cut out, *firm* and accurate creasing of the folds is essential for the production of a neat model. Paper cannot safely be scored with a penknife (as would be done in the case of a cardboard model), but accurate creasing can nevertheless be greatly facilitated by indenting the lines to be creased with a fairly sharp paper-knife.

5.4. It is assumed that the reader knows how to form an ordinary 3-strand plait, such as a small girl uses for her pigtails. Sailors have invented a variety of interesting plaits, including "solid" ones; but we are concerned only with "flat" plaits, being a natural extension of the 3-strand plait to any number of strands. The basic idea is that the strands pass over and under each other in strictly regular order, alternate strands moving the same way. In a true flat plait the strands are turned back each time they reach one or other of the edges of the plait, but in plaiting over a closed surface, such as a poly- hedron, the strands maintain their own directions until they return upon themselves, after which they would continue to encircle the surface indefi- nitely. In designing a plaited model the strands are allowed to overlap their own beginnings just sufficiently to provide adequate "tuck-ins" (see § 5.5).

5.5. Due attention having been paid to §§ 5.3 and 5.4, folding up the model is almost automatic, provided *the first* overlap is correctly made. To this end the section marked O on one strip must be folded over the section marked U on another, the words "over" and "under" being interpreted on the under- standing that the model is being viewed from the outside, i.e. that it is being so held during the course of construction that the outside is uppermost; the over-and-under action of plaiting will then bring each next section to be folded in position as it is needed, and finally one or more ends will remain to be tucked in, which will render the model firm and self-supporting. The nets here shown have been so designed that once the polyhedron is complete, the minimum necessary amount of overlapping is allowed to permit of a finished model in "which the places at which the final tucks are made cannot be determined by visual inspection—in fact a neatly made model can be very difficult to undo! If the first fold is incorrectly made (i.e. if the roles of O and U are interchanged), this will result in the ends of the strips having nowhere to tuck in. The reader wishing to master the art of plaiting polyhedra is advised to start with the simpler ones, where the process will be easy to follow, and work up to the harder ones; the more complex models require a knack which, however, once acquired by practice, enables them to be folded up with surprising rapidity, even when 0 or 8 strands are being simultaneously manipulated. Stellated polyhedra are generally rather tricky, but feasible nevertheless; temporary paper clips are helpful.

5.6. Since the surface of the polyhedron is formed of overlapping strips, and allowance has to be made for final tucking in, the number of faces in the net exceeds twice the number of faces of the polyhedron; moreover the sequence of parts cannot be much varied. Thus only small models can be made with moderate sized pieces of paper; but the method of § 5.8 can be used to overome this.

5.7. Models involving only equilateral or isosceles 120° triangles can most conveniently be made with the help of isometric graph paper, which is ruled into small equilateral triangles (as obtainable, e.g., in the "Chartwell" series, no. 4801—triangles approximately 5 to the inch, effective area of sheet 10″ × 16″). The "Chartwell" paper, which is plain on the reverse side, would in fact be excellent for making models, but as it is rather expensive for cutting up wholesale it is best to use it to determine the vertices of the net or strip, which can then be pricked through on to plain paper.

5.8. A convenient method of preparing the net is to make an accurate construction of a single strip and then prick it through as many times as required—in fact time may be saved by cutting out several together, but to attempt too many at one cutting may result in inaccuracy (a useful dodge, which will reduce the risk, is to staple the sheets together before cutting, being generous with the staples). The strips are then joined together to make the complete net with the aid of small pieces of adhesive paper, and trimmed to the correct lengths as determined from the diagram of the net. This may appear to violate the principle of "no sticking", but it is very quickly and neatly done—the model still being in the flat—and the joins, being inside the finished model, will be unseen. (In theory the strips need not be joined at all, as they would retain their positions in the finished model, but the practical advantages of fixing them in their relative positions at the outset are obvious. Alternatively, the strips could be fastened together by means of slide-on paper clips, which can be removed just before the model is completed; for this purpose the strips must be made long enough to overlap at the outset, rather than joined by edges as they are if stuck. Models which are likely to be undone and refolded at all frequently are however not conveniently made this way, and the firmness of larger models is affected.) This use of separate strips is moreover the only way of making the nets of some polyhedra in which the strips, being curved, cannot be arranged even in the plane net so as not to overlap. It also enables (as mentioned in § 5.6 above) larger models to be made than can be obtained when the whole net has to occupy a single sheet of paper; and if, on these lines, we use different coloured papers for the strips, the results are extremely attractive—and would make fine decorations for the Christmas tree! (I can strongly recommend making in this way the group of models based on **15** (see below), especially the great dodecahedron, **20**. It should be noted, however, that although the colouring does help to bring out the inter-relationships of such a group of polyhedra when viewed in accordance with the ideas of this article, it has no geometrical significance in relation to the individual polyhedron—e.g., it would be more logical to have a great dodecahedron with all five parts of any one plane face rendered in the same colour, whereas when plaited the 12 faces display each of the 6 combinations of 6 different colours chosen out of the 6 (as there are 6 strips), in inverse rotation on opposite faces.

5.9. All construction lines, if any, and joins (if any—as in § 5.8) should be kept on the same side of the net, and the creases can then be made so that this is the inside of the finished model. Several other points of technique which help to increase accuracy and save time could be particularised, but they will doubtless occur naturally to the reader who undertakes to make many of these models.

6. We now come to the practical details of particular models, of which space precludes more than a limited number of examples being dealt with; I try, nevertheless, to include sufficient for the reader to appreciate both the possibilities and the limitations of the method. Note: models marked * are taken (sometimes with slight modification) from Gorham's book; the significance of the index figures will be explained later (see § 7.2).

1. *Cube[3]. Net, Fig. 1. A small model can be very quickly made from ordinary graph paper, as the net can be cut out immediately without even the use of a pencil (although in the classroom it would be advisable to outline the net first). The square A that is finally tucked in goes more easily if slightly tapered, as shown.

2. *Tetrahedron[2]. Net, Fig. 2. The triangles marked x could be dispensed with, but add to the rigidity of the model. The last triangle is a little inclined to untuck itself unless the creases are very firm. (Firmer though less simple plaits for the tetrahedron will be given later—see § 8, 5 and 6.)

3. *Octahedron[1]. Net, Fig. 3.

4. Icosahedron[1]. Net, Fig. 4. Note that two ends have to be tucked in, each consisting of a pair of triangles. This model marked my first success at attempting to extend Gorham's methods.

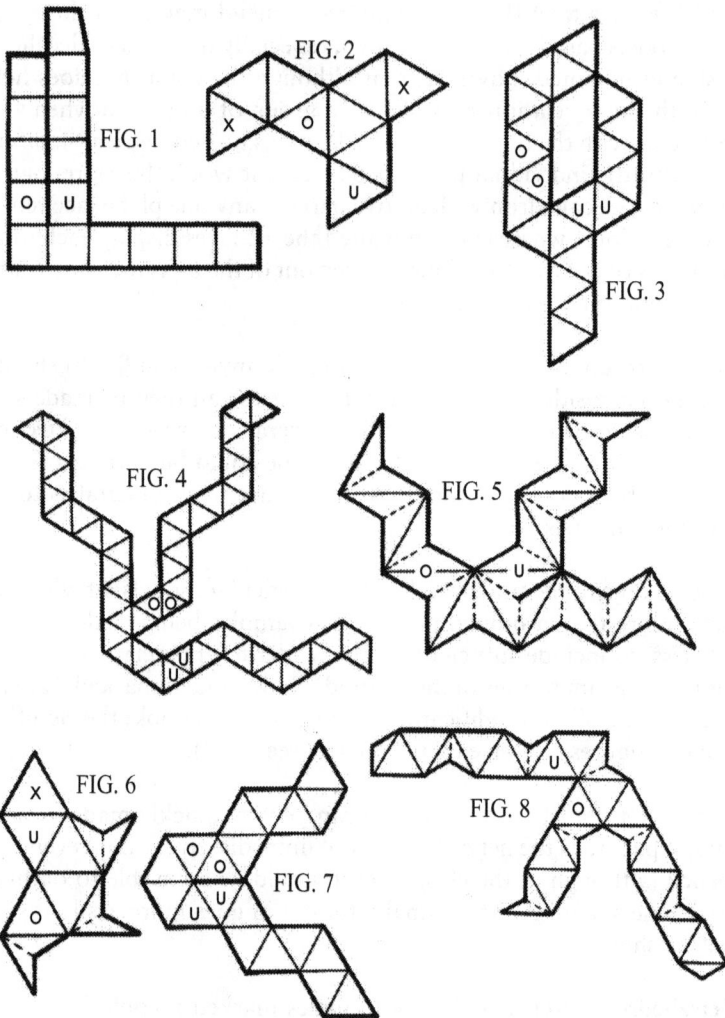

FIG. 1

FIG. 2

FIG. 3

FIG. 4

FIG. 5

FIG. 6

FIG. 7

FIG. 8

A little consideration will show that a dodecahedron, having pentagonal faces, cannot be plaited on the lines of the models so far described. The reader is recommended to make and study the above simple models for himself, and he will then be in a better position to follow the theory underlying the construction of a plaited model of any polyhedron, an account of which follows, and after which I shall resume the description of particular models.

7. General Theory. Where two strips cross, their common region has 4 vertices, and so is a quadrilateral, for there is no loss of generality in supposing the edges of the strips to be straight between the vertices. The quadrilateral may, however, be skew, i.e. folded about a diagonal. The basis of a scheme for plaiting a given polyhedron lies therefore in the dissection of the surface of the polyhedron into quadrilaterals, plane or skew. This can always be achieved by what I shall call *sectoring*: let any point be chosen on each face (but not on its boundary), and joined to all the vertices of that face. Then the desired dissection into quadrilaterals is formed by pairing off those triangles which share an edge of the polyhedron. (Where the faces are regular, it is natural to take the centroid of the face as the "chosen point".) Whilst it is clear that any polyhedron can be dissected into quadrilaterals in this way, the process is generally uneconomical and the number of quadrilaterals can be reduced in a variety of ways, e.g. (1) the polyhedron may already have only (plane) quadrilateral faces—the cube, rhombic dodecahedron and triacontahedron, and trapezoidal icositetrahedron for example; (2) triangular faces adjoining each other may be paired off as they stand; (3) a face adjoined *only* by triangular faces (e.g. a square face of a cuboctahedron) if sectored yields quadrilaterals if the sectors are paired with the complete adjoining triangles; (4) a polygon may be dissected into quadrilaterals and triangles in various ways, and the triangles can be paired off with adjoining triangular faces or sectors.

Having dissected the surface of the polyhedron into quadrilaterals in some way, we now arrange these quadrilaterals into sequences in which the successive quadrilaterals of a sequence are joined by opposite edges. Such a sequence of quadrilaterals forms one of the strips of which the plait is composed.

7.1. The following facts about a sequence so constructed can now be remarked: (*a*) The sequence may cross itself. (This presents no difficulty, it will be found, when the actual scheme for plaiting is developed.) Or, to put it another way, every quadrilateral comes twice, either once each in different sequences or both times in the same sequence. (*b*) Every sequence is closed. For the number of quadrilaterals is finite, and so we are bound to return to the starting point even it we have to cross every quadrilateral in both directions in the process—i.e., there may be only one sequence (and this frequently happens, e.g. in an icosahedron with a suitable pairing of the faces (4)). (*c*) The number of quadrilaterals in a sequence is even. If the sequence is unique, it must cross itself everywhere, and the result is obvious. If not, its edges form the polygonal boundaries of a number of simply-connected regions of the surface. If the number of quadrilaterals in the sequence is odd, some (at least 2, in fact) of these polygons must have an odd number of sides. But a polygon with an odd number of sides cannot be dissected into quadrilaterals (proved inductively by observing that if we subtract from a polygon of n sides a quadrilateral which shares 1, 2 or 3 sides with the polygon then the remaining polygon has $n+2$, n or $n-2$ sides respectively, so that n, if odd, can never be reduced to 4). This contradicts the assumption that the surface of the polyhedron is completely dissected into quadrilaterals. (*d*) Following the edge of a sequence, label the vertices *A, B* alternately. It follows from (*c*) that every vertex of the system of quadrilaterals can be so labelled, in such a way that no two consecutive vertices have the same letter. It is now evident that if we think of the sequences of quadrilaterals as strips (of paper), they can be plaited by following the rule that at each *A* vertex we step "down" from one strip to the one that passes under it if we trace a small contour round the vertex in an anticlockwise sense, whilst at each *B* vertex the same thing happens if we trace such a contour clockwise. This completes the demonstration that any polyhedron can be plaited; and it follows from § 7 that a given polyhedron may be plaited in more than one way.

7.2. If the number of closed sequences on the surface of the polyhedron, i.e. the number of strips needed to form the plait, be *n*, I call the resulting model an *n*-plait. It would appear at first sight impossible to plait with only 1 or 2 strands; this of course is true of a linear (pigtail) plait, but on a closed surface a strand may cross itself repeatedly, and 1-plaits and 2-plaits exist. It is, however, not practicable to *execute* a plait with a single strand or two strands, as this would necessitate forming part of the plait loosely and then threading the rest in and out like a shoelace; to carry out the action of plaiting it is necessary to manipulate simultaneously at least 3 strands (with the exception of certain small models, e.g. octahedron[2] (7), triangular prism[1] (33)), and so in designing the net for (say) a 1-plait it is necessary to cut the strip into at least 3 parts and join them together at intersections—this is what has been done, e.g., for the icosahedron[1] (4). Again, the rhombicuboctahedron and snub cube can most simply be formed as a 3-plait in which one strip is (this is unusual) of different structure from the other two, and this strip is so much longer than the others that it has been broken into three pieces for plaiting so that the net appears to be that of a 5-plait (see 29). The index figures attached to the names of models throughout this article indicate the number of strips of which the model is composed. It should also be made clear that I use the word *strand* rather than strip when referring to the process of plaiting the model rather than the theory of its construction; thus the icosahedron[1] is a plait formed of one strip, but so arranged that in plaiting one manipulates 3 strands. The multiplicity of a plait is clearly brought out by cutting the different strips from different coloured papers; besides being instructive, the results are (as has already been mentioned) extremely attractive.

8. We are now in a position to understand the construction of some further models. But first, returning to those already described, and looking at them in the light of the preceding paragraphs: the construction of the cube is obvious, while the tetrahedron, octahedron, and icosahedron are readily dissected into quadrilaterals by pairing the faces. There are two ways of

doing this for the octahedron, of which the alternative pairing gives a 2-plait described below (7); and the faces of the icosahedron have been so paired as to make it structurally equivalent to a pentagonal trapezohedron. To proceed:

5. *Tetrahedron[3]. Net, Fig. 5. This is a very firm, neat model, obtained by sectoring all 4 faces. (In this, as in some other small models, the end quadrilaterals of the strands have to be truncated as shown or they would be impossible to tuck in.)

6. Tetrahedron[1]. Net, Fig. 6. Only one face is sectored. This is a very useful model, as it is much firmer than 2 but simpler than 5. (The triangle marked × is not strictly part of the plait, but is added for firmness. Similar additions will be found in some of the other small models.)

7. Octahedron[2]. Net, Fig. 7. A different pairing of the faces from that used in 3 leads to this neat 2-plait, in which the two strands are simply "twisted" together to make the model. This is, I consider, easier to make than Gorham's octahedron (3), but care must be taken that it does not fold itself into a triangular dipyramid (see 27).

8. Octahedron[3]. Net, Fig. 8. Two opposite faces sectored, giving a very firm, neat, yet fairly simple model.

9. *Octahedron[4]. Net, Fig. 9. In this and the next model, all faces are sectored.

10. *Cube[4]. Net, Fig. 10. See note to 9.

11. *Rhombic dodecahedron[4]. Net, Fig. 11.

8.1. The relationship between the last three models merits careful study. In 10 the faces of the cube are divided into right-angled isosceles triangles, and the basic quadrilaterals of the strips are squares folded each about a diagonal, but not along their common edges. Now suppose that these squares are replaced by congruent rhombuses with acute angles at those vertices corresponding to the uncreased diagonals, and that additional creases are made along their common edges; then on plaiting the model, a pyramid will be raised on each face of the cube. (In particular, if we take the above mentioned acute angles to be 83° 37′ we shall obtain a

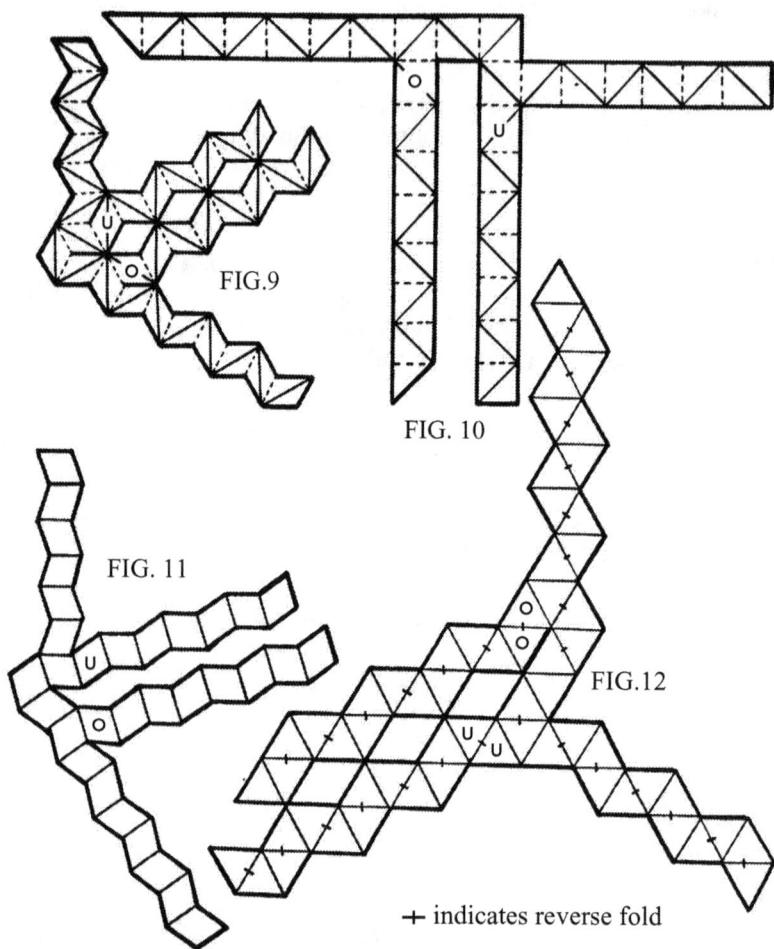

FIG.9

FIG. 10

FIG. 11

FIG.12

+ indicates reverse fold

12. Tetrakis hexahedron[4]. But if the additional creases are reversed, the faces of the cube will acquire pyramidal "dimples".) Suppose now that we raise pyramids of such a height that the adjoining faces of two adjacent pyramids lie in the same plane; we then have a solid with as many rhombic faces as the cube has edges, namely a rhombic dodecahedron: the diagonal creases can now be dispensed with, and we have model 11. (The acute angle of the rhombus for this purpose is 70° 32′, but it is perhaps easier to make use of the fact that its diagonals are in the ratio $\sqrt{2}$: 1.) In the same way, starting with an octahedron plaited as 9, on replacing the 120° angles of the rhombuses of which the strips are composed by 117° 14′ and creasing along the common edges we obtain a

13. Triakis octahedron[4]; and by making them 109° 28′ (the supplement, of course, of the 70° 32′ mentioned above) and omitting the diagonal creases we obtain again the rhombic dodecahedron 11. Further, by making the same angles 60° and reversing the diagonal creases a tetrahedron will be raised on each face of the basic octahedron and we obtain a

14. Stella octangula[4]. Net, Fig. 12. (This is rather tricky to fold up.) Thus we see that all the models 9 to 14 are formed from basically the same plait, and the similarity of the nets is evident. I use the term *compound plait* to designate a plait formed by sectoring all the faces of a polyhedron, e.g. 5, 9 and 10. It is now clear that we have available a system whereby a range of related polyhedra can be formed from one basic net by altering the angles of the constituent quadrilaterals and adjusting the position and "sense" of the creases in conformity. (In particular it should be remarked that the compound plaits of any convex polyhedron and its dual differ only in (*a*) the angles of the quadrilaterals and (*b*) the transference of the crease from one diagonal to the other.) In this way, an interesting group of models can be based upon the compound plait of the icosahedron, as follows:

15. Icosahedron[6]. Net, Fig. 13. The basic quadrilaterals are 60°, 120° rhombuses, creased along the long diagonals.

16. Triakis icosahedron[6]. In 15, replace the obtuse angles by 119° 3′, and crease additionally along the common edges.

17. Rhombic triacontahedron[6]. In 15, make the obtuse angles 116° 34′ and crease along the common edges only.

18. Pentakis dodecahedron[6]. In 15, make the obtuse angles 111° 24′ and crease along the common edges and the short diagonals.

19. Dodecahedron[6]. In 15, make the obtuse angles 108° and crease along the short diagonals only. This is the compound plait of a dodecahedron, and its relationship to that of the icosahedron via the rhombic triacontahedron is the same as that of the cube to the octahedron via the rhombic dodecahedron. No substantially simpler plait for the dodecahedron appears to be possible, which is a pity as it means that this only of the 5 Platonic solids cannot easily be made in the classroom by the present method.

20. Great dodecahedron[6]. This solid may be regarded as an icosahedron with dimples, and it is easy to discover that its net will have the same angles as 19 but that the creases must be made on the long diagonals of the rhombuses and, reversed, on the common edges. The complete identity, apart from the creases, of the nets of the dodecahedron and great dodecahedron is unexpected and striking. This model, which is not at all difficult to fold up, is one of the firmest and most satisfying that I have made by this method.

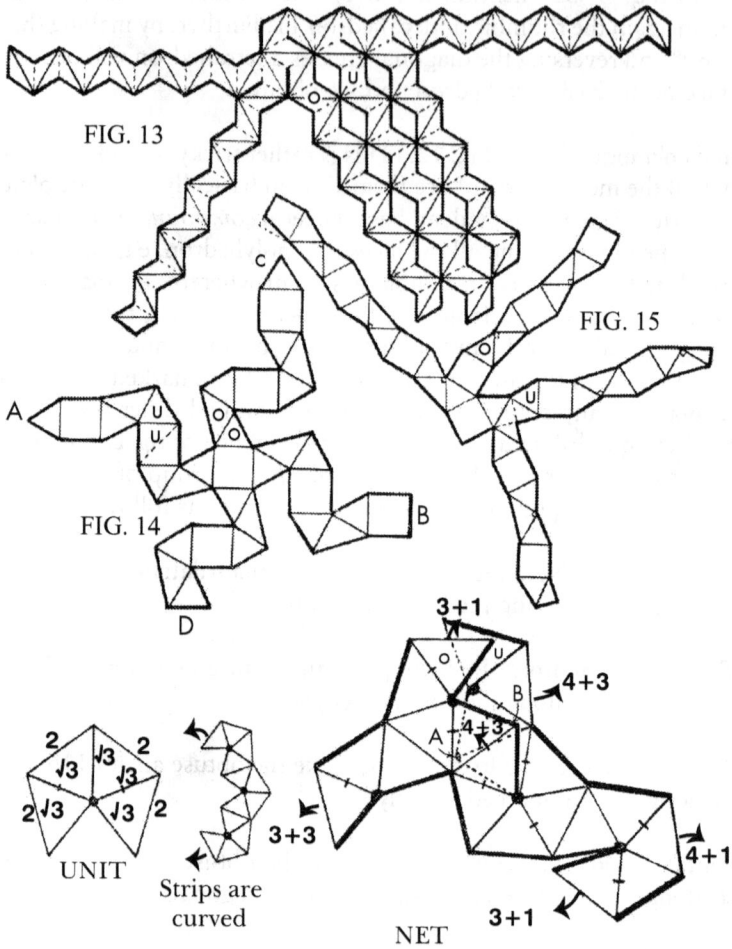

FIG. 13

FIG. 15

A

FIG. 14

B

D

UNIT

$$2 \quad \sqrt{3} \quad 2$$
$$2 \sqrt{3} \quad \sqrt{3} \quad 2$$

Strips are
curved

3+1

4+3

B

A **4+3**

3+3

NET

4+1

3+1

FIG. 16

Arrows show <u>further</u>
number of whole units
+ extra triangles at end

21. Small stellated dodecahedron[6]. Regarding this as a dodecahedron with pentagonal pyramids raised on its faces, the net can be deduced from that of 19 in a way that should now be clear to the reader.

22. Great stellated dodecahedron[6]. Regarding this as an icosahedron with triangular pyramids raised on its faces, its net can be deduced from that of 15.

Stellated polyhedra are rather tricky to fold up, but both of the last two models have been successfully made by my pupil Mr. R.W. Bray. It is easily seen (v. the illustrations in *Mathematical Models,* pp. 84 and 90) that the great icosahedron may be regarded as a small stellated dodecahedron with dimples; its net can thus be deduced from that of the compound plait of the latter. Mr. J.C. F. Fair has tackled the plaiting of this model with success; the strips turn out to be "straight" and of comparatively simple structure.

8.2. Readers who are sufficiently interested to try their own hand at making models by the method of this article will find that convex polyhedra are fairly easy to fold up; after a floppy start, the strands gradually but surely shorten under one's fingers as one vertex after another is completed, and the model acquires firmness in a rather sudden and very satisfying manner as the stage of tucking in the ends is reached. (The number of ends to be tucked in is about half the number of strands—which, it should be remembered, may exceed the number of strips: cf. § 7.2.) Stellated polyhedra are more difficult to fold up, although the knack can be acquired, and the resulting models are less firm—though quite satisfactory in appearance; but "dimpled" polyhedra, such as the great dodecahedron, make very firm models.

To continue with some further models of interest:

23. **Cuboctahedron**[2]. Net, Fig. 14. This is the simplest plait of this solid; the two strips are AB and CD, which, being almost identical, can be cut out together, and then plaited without initial fixing, AB being placed over CD.

24. **Cuboctahedron**[4]. Net, Fig. 15. This is a rather neater model than 23.

25. **Cuboctahedron**[6]. This is the compound plait of the solid, and it is hardly worth making on its own account, but from it can be deduced the net of the compound plait of its dual, the rhombic dodecahedron (or of course this can be worked out independently), and from either that of the

26. **Stellated rhombic dodecahedron**[6]. Net, Fig. 16. I have made a successful model of this. When folding, note that at the start the corners marked A and B are brought together to form an outward-pointing vertex.

Deltahedra are polyhedra all of whose faces are equilateral triangles (see *Mathematical Models,* pp. 72 and 135f.). They include, of course, some of the regular polyhedra, but I have also made several of the others, both convex and non-convex; the irregular ones need folding with care, as they have a disconcerting tendency to fold up into the wrong deltahedron. A very neat model is the

27. **Triangular dipyramid**[1]. Net, Fig. 17. This net is a shortened version of that of the octahedron[2] (7).

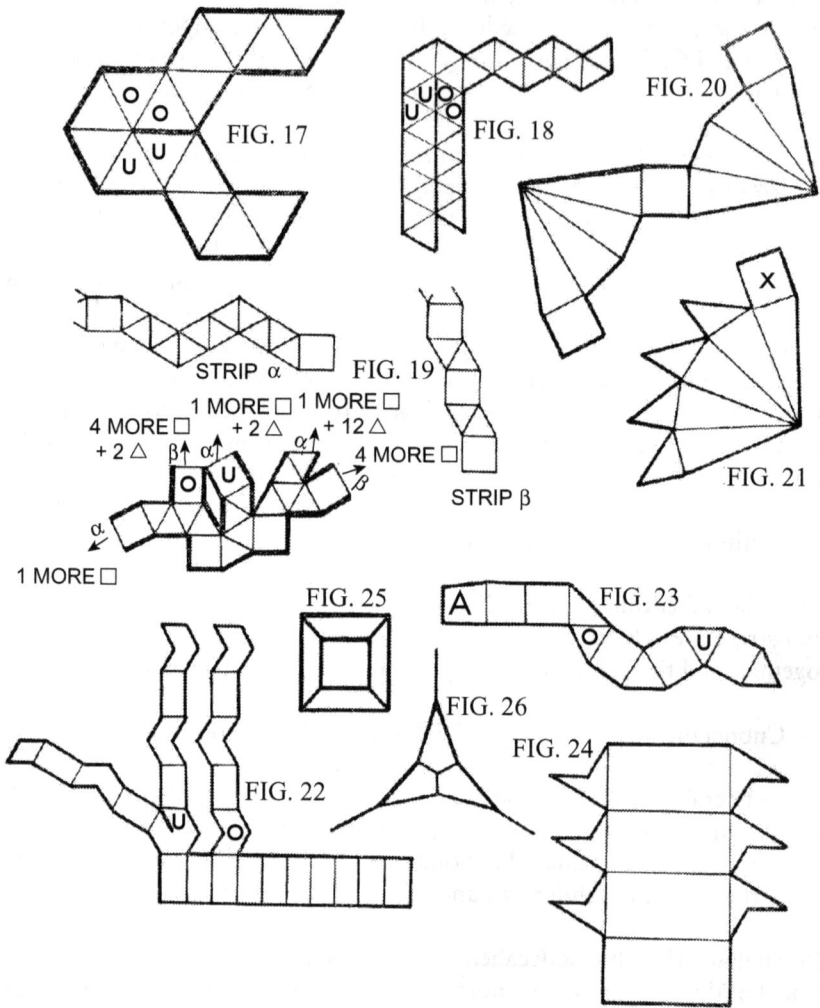

FIG. 17

FIG. 18

FIG. 20

STRIP α   FIG. 19

FIG. 21

1 MORE □ 1 MORE □
4 MORE □   + 2 △   + 12 △
+ 2 △   β   α   α   4 MORE □
              O   U   β

α   STRIP β

1 MORE □

FIG. 25   A   FIG. 23

FIG. 26

FIG. 24

FIG. 22

Another example is the

28. Dodecadeltahedron[3]. Net, Fig. 18.

If the "cube" faces of a rhombicuboctahedron are left alone, and all the other square faces divided by diagonals (which must be suitably chosen), then on replacing all the right-angled isosceles triangles so obtained by equilateral triangles we get a snub cube. The nets of these two solids will therefore be closely related. The snub cube can be dissected into quadrilaterals by suitable pairing of the triangles (not, of course, that corresponding to the rhombicuboctahedron, which would leave odd ones over), and the resulting sequences give two short strips and one long one of different structure, so that it is in fact a 3-plait; but the net has the appearance of a 5-plait as the long strip has to be broken into 3 pieces to make the model workable. I illustrate the snub cube, as being the more interesting of the two solids. The enantiomorph should present no difficulty. Alternatively, undo the one given, reverse the creases, and refold—plaited models can be turned inside out in *three* dimensions!

29. Snub cube[3]. Net, Fig. 19.

8.3. There is of course no need for the polyhedra to have regular faces for them to be successfully plaited: in fact a large part of Gorham's book is devoted to forms which the geometer would regard as irregular but which occur in the study of crystals. The reader of this article should have no difficulty in modifying the nets of (e.g.) the tetrahedron and cube to be able to plait scalene tetrahedra, cuboids and parallelepipeds.

There are a few other solids of frequent occurrence whose nets I now give, some of the forms (those with no index) being ad hoc constructions not strictly falling under the general theory.

30. Square pyramid[1]. Net, Fig. 20. This is of interest as being a 1-plait in which the single strand can be folded up as it stands. The next model is however simpler and neater.

31. Square pyramid. Net, Fig. 21. The square marked × is added for firmness. A resemblance will be noted to the tetrahedron plaited as in 6. The idea can be successfully extended to pyramids of any number of sides.

32. *Hexagonal prism[4]. Net, Fig. 22. A hexagonal prism with the ends divided into three rhombuses is structurally equivalent to a rhombic dodeca hedron (with some creases omitted), but I give the net here separately as a model of this solid is likely to be in demand. To make a bee's cell, replace one end of the prism by a set of 3 rhombic dodecahedron faces, and modify the rectangular faces accordingly (using the same angles as for the rhombuses).

33. Triangular prism[1]. Net, Fig. 23. Another 1-plait that can be folded up as it stands. The part A should strictly be a triangle, but a rectangle (tapered for ease in tucking-in) is less liable to come undone.

34. Triangular prism. Net, Fig. 24. This is an *ad hoc* construction which makes a neat model.

Gorham's own version of the triangular prism I do not give, as it is cumbersome compared with either of the above.

8.4. Amongst other models which I have made by this method are the great dodecadodecahedron, and the (space-filling) truncated cuboctahedron. In *Multi-Sensory Aids* the expression *intraverted polyhedron* is used to describe the pseudo-solid obtained by joining the edges of a polyhedron by means of planes to its centre, and removing the original faces—thus each such face is replaced by a pyramidal dimple, all the pyramids having a common vertex. An *extraverted polyhedron* is the same solid with the pyramids reversed on to the outside of the figure. Using this terminology, it is well known that an extraverted cube is a rhombic dodecahedron; hence an intraverted cube is easily plaited from the net of the rhombic dodecahedron by creasing along the short diagonals and reversing the original creases. To make an intraverted tetrahedron, take the net of the tetrahedron[3] (5) and make the acute angle of the rhombuses 70° 30'.

9. The designing of a net for plaiting a desired polyhedron—if not deducible from one already known by methods such as those described in § 8.1—is greatly facilitated by the use of a "distorted" or Schlegel diagram. There are two versions of these: (a) in which one face of the polyhedron is removed and the remainder of the surface is shrunk and flattened so as to lie within the boundary of that face (which it is now convenient to think of as represented by the rest of the plane), and (b) in which the polyhedron is similarly flattened but one vertex is projected to infinity. Without troubling to make these descriptions more precise, the processes will become clear on referring to Figs. 25 and 26, which show the results of (a) and (b) for a cube.

If the Schlegel diagram is sketched in pencil, the dissection into quadrilaterals can be done in ink, and then the sequences of quadrilaterals can be followed by means of differently coloured lines, and the actual forms of the strips then worked out. It should be clear that there is no choice in the matter of the sequences—they are determined completely by the dissection into quadrilaterals, but this, as already pointed out, can be done in numerous ways.

The simplest procedure, of course, is to trace out the strips on an actual (not necessarily plaited) model, if one is available!

10. In one of the earlier attempts at experiment in the teaching of geometry to beginners, by G.C. and W.H. Young [5], the authors encourage throughout the making of models, which they have designed so that they can be folded up and secured by means of tucked-in or interlocking flaps without sticking. The net so prepared they call a "flat pattern", and consists of the conventional net with the addition of sufficient extra faces to enable a self-supporting model to be formed on similar lines to those of this article. For so simple a solid as the regular tetrahedron it is not surprising to find that their flat pattern is identical with the net for plaiting as given by Gorham (which is 2 above without the triangles marked ×), but for the cube and all other models the positioning of the additional faces in the flat pattern,

although sufficient for the purpose, is otherwise quite arbitrary, and the models cannot be readily folded up without a knowledge of the exact order of interlocking of the faces. To this end the faces in the flat pattern are numbered and lettered systematically: Gorham does the same thing, but in the case of his models it is unnecessary, with the exception of the first fold, as the beauty of plaiting is that provided this is made correctly the remainder of the process of folding, however large the model, is automatic. The complications of plaiting (say) a dodecahedron do not arise in the Youngs' book, as they do not go beyond the octahedron and some simple irregular solids (including Bimbo's Lozenge!—a deltahedron in the form of a pentagonal dipyramid, which incidentally they incorrectly describe as being formed from five regular tetrahedra. This can be made by extending the net of 7 in the same way that the latter can be obtained by extending the net of 27.) The Youngs' models are less firm than plaited ones, as there are more "ends". An interesting feature of the flat patterns, however, is that they are all obtained by folding a plain sheet of paper; no construction lines are used, the only apparatus needed being a sheet of paper and a pair of scissors. (I wonder how many readers could fold an equilateral triangle?) This idea can quite well be applied to some of the simpler plaited models; it requires skill, and care not to introduce unwanted creases in plane faces, but can save time when mastered.

11. It is surprising that the idea of plaiting a cube, tetrahedron, and octahedron by the simplest method (1,2,7) is not more generally known, and possibly there have been others besides Gorham and the Youngs who have struck out on these lines; but to Gorham must go the credit of making these solids by what I have termed a compound plait. I do not know how he hit upon this method, although presumably it arose from the nature of the crystal structures in which he was interested; but it was a brilliant and fruitful inspiration, without which this article would not have come to be written.

REFERENCES

1. H.M. Cundy and A.P. Rollett, *Mathematical Models,* Oxford, 1952. Reprinted by Tarquin, 2006.

2. *Multi-Sensory Aids in the Teaching of Mathematics* (18th Yearbook of the National Council of Teachers of Mathematics), Columbia University, New York, 1945.

3. John Gorham, M.R.C.S., *A System for the Construction of Plaited Crystal Models on the Type of an Ordinary Plait; Exemplified by the Forms Belonging to the Six Axial Systems in Crystallography,* E. and F.N. Spon, London, 1888.

4. W.W. Rouse Ball, rev. H.S. M. Coxeter, *Mathematical Recreations and Essays,* Macmillan, 1940.

5. G.C. and W.H. Young, *The First Book of Geometry,* Dent, 1905.

THE PLAITED DODECAHEDRON

By JAMES BRUNTON

Mr. Pargeter deserves our thanks for introducing us to the fascinating art of plaiting polyhedra (May 1959) and I would like to indulge in the luxury of two comments on what he said.

(i) In para. 5.7 Mr. Pargeter says, "Chartwell Paper is excellent for aiding the construction of nets using only 60° or 120° etc." but I use a simpler method, which has the advantages of being much cheaper and of being something that the pupil can prepare himself. The paper folding in the following diagrams is very accurate if carefully done and provides some good revision for the abler pupils. Solid lines indicate the edges of the paper and thin lines the folds. Folds 2 & 3 are then repeated from the right-hand side to complete this basic triangle and it can then be sub-divided into an all-over pattern of smaller ones, either by folding A & B on to the centroid G and then parallel folds, or into even smaller triangles by folding the height into half and then quarters. The completed pattern is used by my pupils as the template for nets in the way Mr. Pargeter suggests, by pricking through on to plain paper, or else, as we sometimes do, by direct cutting.

(ii) In his note on model 19, after indicating how the compound plait for the dodecahedron can be developed, Mr. Pargeter says "No substantially simpler plait for the dodecahedron appears to be possible, which is a pity as it means that this only of the 5 Platonic solids cannot easily be made in the classroom by the present method."

May I submit the following suggestion which is in the spirit of Mr. Pargeter's 3-plaits if not in the letter. In his note to model 4 he says that having pentagonal faces the dodecahedron cannot be plaited on the lines of the other models. But ten of the twelve faces can! and in the present model the other two faces, a parallel pair, are by the *ad hoc* method suggested for the base of the pyramid and ends of the prism, models 31 & 34. Without a complicated means of preliminary folding, the three strips are best cut separately. They are similar in shape and are cut according to Fig. 5, strip No. 1 being A to N, strip No. 2 A to L and strip No. 3 D to O. $3O$ is now stuck

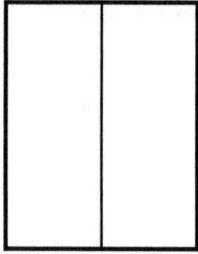

Fig. 1

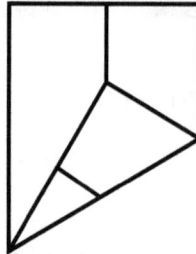

Fig. 2

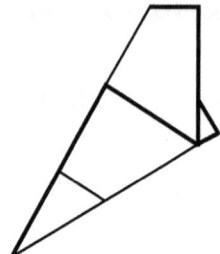

Fig. 3

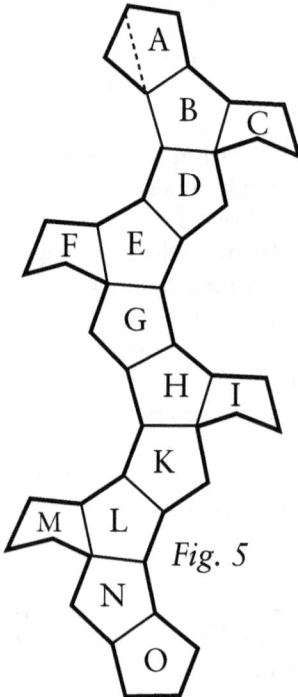

Fig. 5

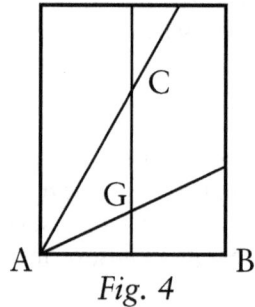

Fig. 4

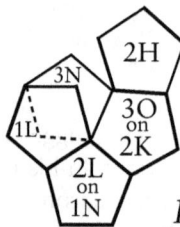

Fig. 6

Figure 1.   The first fold is the perpendicular bisector of the base-line.

Figure 2.   This fold is best done by keeping the thumb-nail (nail-biters cannot become good mathematicians!) on the left-hand corner and stretching the base edge of the paper with the right hand while folding.

Figure 3.   This is a reversed fold for accuracy.

Figure 4.   This shows the first half of the process completed and it is easy to see that the $\Delta\ ABC$ will be equilateral.

down to $2K$ and $2L$ to $1N$ as shown in Fig. 6. Plaiting then begins by placing $1L$ over $3N$, then $2H$ over $1K$ etc. $1A$ tucks under $1L$ at last, and it is best cut and folded as indicated as it is then less likely to slip. Finally the remaining two faces are formed by tucking the flaps under each other in rotation. This model is firm and as quickly plaited as the other four Platonic solids suggested by Mr. Pargeter.

I encourage as much folded-paper Geometry as I can, since many constructions are more quickly and more logically (and more accurately!) performed and when Mr. Pargeter challenges us to fold an equilateral triangle, I suppose he is thinking of some such method as I referred to above. I would (and do) also use folded pentagons for the development of the dodecahedron.

Chatham Technical School for Boys   JAMES BRUNTON

Editorial Note. Mr. A.R. Pargeter made the following invited comment on Mr. Brunton's construction:

I am most interested to see Mr. Brunton's 3-plait for the dodecahedron. When I was first experimenting at extending Gorham's methods, and before I had fully grasped the principles that I set forth in my article, I did make a dodecahedron on a practically identical plan, but I rejected it on account of its lack of symmetry and the number of "ends"; in any case, it was clear that one could not plait a dodecahedron with pentagons only—free vertices would be left flapping, as it were—and I felt sure that there must be something analogous to Gorham's "compound" octahedron. In Mr. Brunton's arrangement the free vertices are all inside the model; and it may in fact be regarded as an ingenious extension of the method I used to construct a triangular prism, model **34** in my article. As beginners find a 6-plait rather tricky to handle, Mr. Brunton's model may well help to fill the gap that I lamented, namely that in the classroom the dodecahedron alone of the 5 Platonic solids was not easily made by plaiting.

A SYSTEM

FOR

THE CONSTRUCTION

OF

CRYSTAL MODELS

ON THE TYPE OF AN ORDINARY PLAIT;

EXEMPLIFIED BY THE FORMS BELONGING TO THE SIX AXIAL SYSTEMS IN CRYSTALLOGRAPHY.

E. & F. N. SPON, 125, STRAND, LONDON.

NEW YORK: 12, CORTLANDT STREET.

1888.

PREFACE

IT is now some forty years since I had the honour of demonstrating before the Royal Society in London *"A System for the Construction of Crystal Models projected on Plane Surfaces."* These figures folded into the required form, and subsided into a level at pleasure—they were easily moulded into shape by bringing their edges into apposition with the fingers, and were as easily transferred from place to place when flattened in a portfolio—they constituted, in short, an extension of the plan used in modern mathematical treatises for extemporising models of the five regular or Platonic solids. At that period, besides diagrams in isometric perspective, which are as necessary now as they were then, models in wood were much in vogue. They were used by Dr. Pareira in his lectures on the Polarisation of Light before the Pharmaceutical Society of Great Britain. Phillips, in the second edition of his 'Introduction to Mineralogy, recommends the use of models cut in box-wood, and observes that "they could be had at one guinea each, while complete sets could be procured at the price of £16 the set" ; at this rate the study of crystallography became somewhat expensive. In modern times it has been found possible to arrange the many thousand different known crystals in *six systems,* to each of which belongs a number of forms having some property in common. Each system consists of a skeleton model of three or four rods of wood, wire, or glass; these rods are called axes, round which the forms can be symmetrically built up. Upon these axes it is proposed, in the first place, to find the faces of the required model by direct measurement (or recourse may be had to Spherical Trigonometry, as the case may be), and in the next place, to build them up into a model by a process which it is believed has not been hitherto attempted. It consists in taking an ordinary plait of three or four rushes, defining its intersections in numerical order, and thus eliminating the *type* on which every model is constructed. By strictly adhering to the type it was found, moreover, that those solids which were confessedly irregular and most difficult to understand—those, for instance, belonging to the *doubly oblique system*—were made with the same facility as the cube itself.

Considering how the zones in crystals approximate in appearance to the ideal plait, it is singular that this method of making models should so long have escaped notice. The first intimation of the possibility of forming a crystal model in this way suggested itself in the case of the *rhombic dodecahedron,* a solid bounded by twelve rhombs, and characterised by exceeding elegance of form—some of the smaller specimens of garnet constituting the most aesthetic "things of beauty" in all mineralogy. On a careful examination of one of these crystals, its faces appeared to be arranged in narrow strips, which could be traced round the form in four different directions, and seemed to cut each other in their course as if they intersected. It became difficult not to realise the practicability of using strips of paper—of crossing them just as these zones appeared to be crossed in a real crystal, and of inter-twining them as in a plait. Four narrow strips of paper were taken accordingly, each being composed of similar rhombs placed together at their opposite edges, after crossing; and recrossing repeatedly, a rough model of the form was eventually obtained. The ability to make a single solitary model, however, and that by a mere hap-hazard, was not altogether encouraging. On reconsidering the matter, it became evident that the definite arrangement of the parts in a plait could be at once utilised by finding the numerical order in which its intersections occurred. This formed a clue to the whole. The idea, when practically worked out, fulfilled my most sanguine expectations, as shown by the results which are recorded in the following pages.

Compared with others, the advantages resulting from this method become most strikingly apparent. The models are built up into form in a few seconds, and it is worthy of notice that, owing to the plaiting process being well nigh instinctive, the manipulations after a short trial become almost automatic; the eyes may be closed—the attention diverted—yet the required solid meanwhile may be growing into existence. The forms require no sticking at their edges, being far neater than those which do. They have the semblance of solidity, and as each face involves at least two thicknesses, their stability is secured. They allow of the measurement of their edges with a *goniometer.* Being so easily made, an opportunity is afforded of enclosing a series of gradually enlarging hollow models (see Figs. 97, 101, 104) over each other upon a central nucleus (like nested boxes)—the process imitates the growth of a natural crystal, and the compound nucleated model indicates by the use of needles the direction of its *cleavage.*

Finally, the plaiting of a cube suggests a possible manner in which the crystal particles may start originally from a common root, and travel in certain definite directions until the material by which they are fed becoming exhausted, their growth ceases, and their form is completed.

JOHN GORHAM, M.E.C.S. Eng., &c.
Tonbridge, Kent.

A SYSTEM FOR THE CONSTRUCTION OF CRYSTAL MODELS ON THE TYPE OF AN ORDINARY PLAIT

It is a property inherent in the ordinary plait that its constituent parts shall cohere compactly without adventitious aid. This involves definite arrangement; and if this arrangement is studiously followed in the construction of models, their faces will in like manner become coherent, and, when plaited, will assume a solid form, and maintain it without extraneous assistance. Hence, to make a model by this method, it is essential that the natural collocation of the parts in an ordinary plait shall be defined in numerical order, and that this order, when once obtained, shall be implicitly followed in the making of every crystal form, however much the shapes of its faces may vary.

An ordinary plait of three or four flexible narrow strips is so easily put together that the process would appear scarcely more than intuitive. On making an outline of the common plait, Fig. 1, we notice that the strips intersect or cross one another in a manner so definite, that it may be easily specified in numbers, each number, from below upwards, indicating the place of intersection. That part of the one strip which overlaps is denoted by an ordinary numeral; that which is overlapped, or lies underneath, by the same numeral, barred thus: 1 and $\bar{1}$. On now unfolding the form we find the projection, Fig. 2, with its numbers so disposed, that, were we previously unable to make this ordinary plait, the projection alone would enable us to do so. It will be noticed, however, that the figure thus obtained is capable of extension in one direction only: it may be elongated, but is devoid of solidity. To obtain a model having the form of a *solid,* we take three strips composed of square faces, Fig. 3, upon which we transcribe the numbers exactly as they are disposed in Fig. 2; and finally, instead of allowing the three strips to lie parallel on commencing the plaiting, we place the strip marked C at right angles to the other two strips marked A and B. The projection, Fig. 4, will now make up into the *cube,* Fig. 5, shown in its projected state, Fig. 6.*

The numerical order in which the faces combine to form a plaited model of the *cube* is thus secured, and this order is retained as the type on which crystal models may be constructed.

---

* The best material for making these models has been found to consist of *glazed cambric* and *white demy paper,* pasted together into sheets, and well pressed, so as to be entirely clear of air-bubbles. Cardboard is useless, breaks at the bends, and drops to pieces. On making a projection of the pattern on a flat surface, a face of the model is cut out in cardboard : a square for the cube, for instance, round which the outline is traced with a pencil; for if drawn with ruler and compasses, the faces are brought too close to plait

The uses to which the *cube* is applied are so multifarious, that the ability to construct a model of it by a process at once so easy, certain, and rapid, must needs be a desideratum. Amongst the natural crystal forms a very considerable number are geometrically allied to the cube, and the common type on which their configuration is based is so absolutely identical, that the primitive forms belonging to the six axial systems in crystallography may, one and all, be constructed upon it. Take, for instance, the parallelepipeds known as the six-sided *prisms;* the *rhombohedrons,* acute and obtuse; the *octahedrons,* from the regular to the doubly oblique; the elegant *scalenohedron* with its allied *calcite.* By strictly adhering to the numerical order in this cubic projection a series of *sub-types* can be produced, resulting in modifications which exemplify the forms of natural crystals; witness the *typical tetrahedron,* Fig. 24, under which head are comprised those irregular tetrahedral solids named *sphenoids,* themselves being *hemihedral* forms of their respective octahedron.

Owing to the remaining forms being contained under a greater number of faces, they cannot, of course, be made by using a cube composed of three fillets only. An additional fillet is therefore added, by the aid of which we are enabled to construct a *second* cube more composite in its nature, but still a cube, the projections of which supply us with all the remaining solids. The cubic type thus maintains its supremacy, while the unity of the system is preserved.

---

properly. It imparts finish to a form by cutting separate pieces from a sheet of stone-coloured cardboard, exactly equal in number and shape to the faces of the model itself. When pasted with care, the form becomes scarcely distinguishable from that of a solid. Sample specimens may be procured, however, of Mr. Henson, Mineralogist, 227, Strand, London, W.C.

Plate 1

Fig. 1.

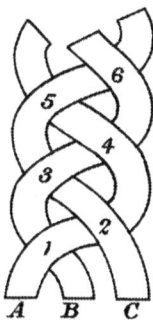

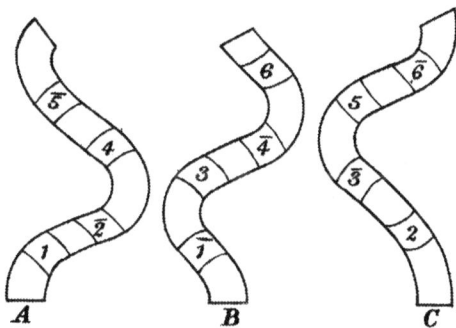

Fig. 2.

E & F.N.Spon, London & New York

Plate 2

Fig. 3.

A	B	C
		$\bar{\bar{6}}$
$\bar{5}$	6	5
4	$\bar{4}$	$\bar{3}$
$\bar{2}$	3	2
1	$\bar{1}$	0
0	0	0

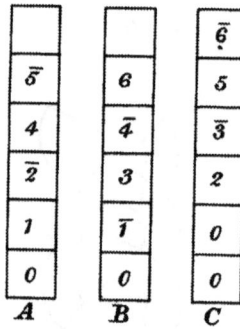

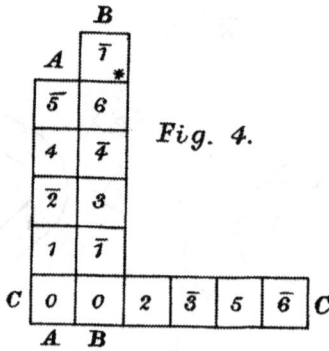

Fig. 4.

Fig. 5.

Plate 3

CUBE

ADJUSTMENT OF ITS PLANES DERIVED FROM ORDINARY PLAIT.

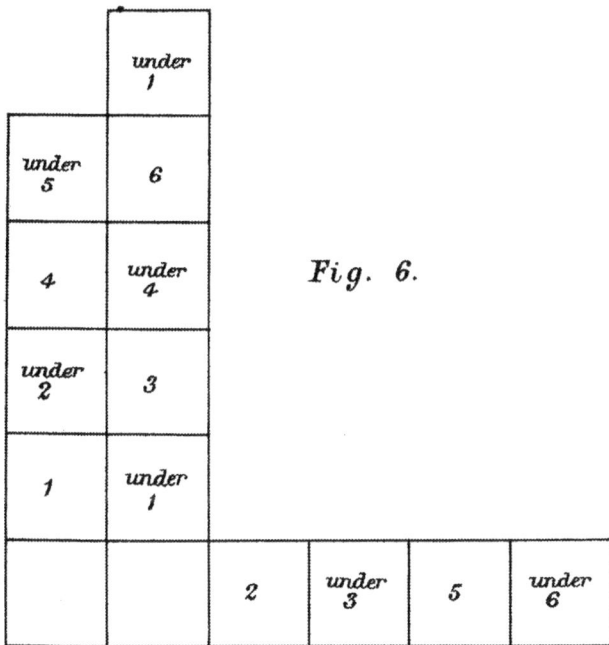

	under 1
under 5	6
4	*under* 4
under 2	3
1	*under* 1

Fig. 6.

		2	*under* 3	5	*under* 6

E & F.N.Spon, London & New York

Plate 4

Fig. 8 the same unplaited. The spaces overlapped in each strip are distinguished by barred numbers thus $\bar{7}$ which means under 1. &c.

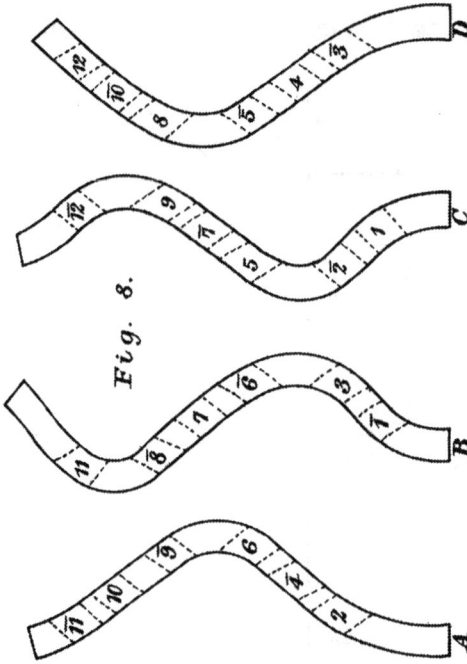

Fig. 8.

Fig. 7. Plait of 4 strips.

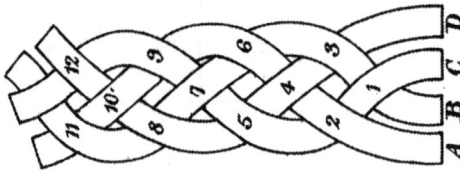

Fig. 7.

E & F.N.Spon, London & New York

To construct a *cube* with a plait of *four* fillets, let A B C D, Fig. 7, be four strips intertwined in the usual way to form a plait, and let them be numbered in the order of their intersections, from below upwards, as in the diagram. Then, as each intersection involves two faces, the one overlapping the other, let the upper one be denoted by an ordinary numeral, say 3, and the lower one with a *barred* numeral, say 3̄, as in Fig. 8.

Fig. 9.—The strips are drawn in a rectilinear direction. They are composed of square faces, which are numbered like those in Fig. 7. The base planes are indicated by small circles (O); those which give security to the whole are defined by an asterisk (*).

In Fig. 10, the four series are arranged in the true order for plaiting, in which the two innermost strips, B and C, are placed perpendicular; while A and D are projected horizontally. A diagonal in dotted lines across each square completes the figure, and each square is creased along the diagonal.

The process of plaiting commences by bringing the face numbered 1 in C over that marked 1̄ in B; next, the face numbered 2 in A is carried over that marked 2̄ in C, and so on until the planes are exhausted.

Fig. 11 shows in isometric perspective a model of the resulting solid. It is a cube, each face of which, instead of being one entire piece, is composed of four isosceles triangles of 90°, which appear set like a mosaic. It has been constructed in this way for the purpose of showing many of the highly interesting forms which are recognised in crystallography as modifications on the faces of a cube. These modifications are produced by varying the angular value of the isosceles faces. When they are equal to 90°, the form is a cube. When less than 90°, say 83°, a pyramid of four sides is raised on each face as in some varieties of *quartz;* while, if still further reduced, say to 70° 31′ 44″, the resulting model is that of the wellknown and beautiful form known as the *rhombic dodecahedron,* of which the garnet is a familiar example.

The Rhombic Dodecahedron is bounded by twelve similar and equal rhombs, the plane angles of which are equal to 109° 28′ 16″ and 70° 31′ 44″.

The faces incline to each other at the edges at an angle of 120°. It is geometrically allied to the cube, for if the twelve edges of the cube are replaced by tangent planes, and these are extended until they mutually intersect, the rhombic dodecahedron will be formed.

To find the *plane angles of the rhombic face*:— These are discovered by the formula $L\dfrac{1}{2\ \sin\frac{1}{2}A} = \cos\frac{1}{2}a,$

where $\cos\frac{1}{2}a = 54°\ 44'\ 8''$

and $\qquad a = 109°\ 28'\ 16''*$

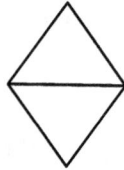

Rhomb and Short Diagonal.—We now revert to our typical projection, Fig. 10, with its *zigzag dotted diagonals,* and trace on a sheet of paper outlines of the rhombic face in the same numerical order, making its *short diagonal* correspond in direction with that of the *zigzag* lines. The faces are now numbered in the same way, and the projection, Fig. 12, will plait into an elegant and firm model of the *rhombic dodecahedron.* (Fig. 13.)

---

* A mode of finding this plane, which is useful for practical purposes, is to make the short diagonal of the rhomb equal to the side of a square, and the long diagonal of the rhomb equal to a diagonal of the square. A more elegant mode is as follows:—Let the short diagonal of the rhomb be taken as 1, then the long diagonal will be equal to, $\sqrt{2} = 1.414$ and any side of the rhomb will be equal to

$$\frac{\sqrt{3}}{2} = \frac{1.732}{2} = 0.866$$

Plate 5

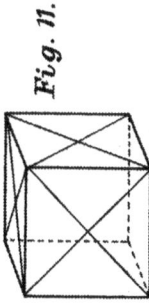

Fig. 11.

TYPICAL CUBE.

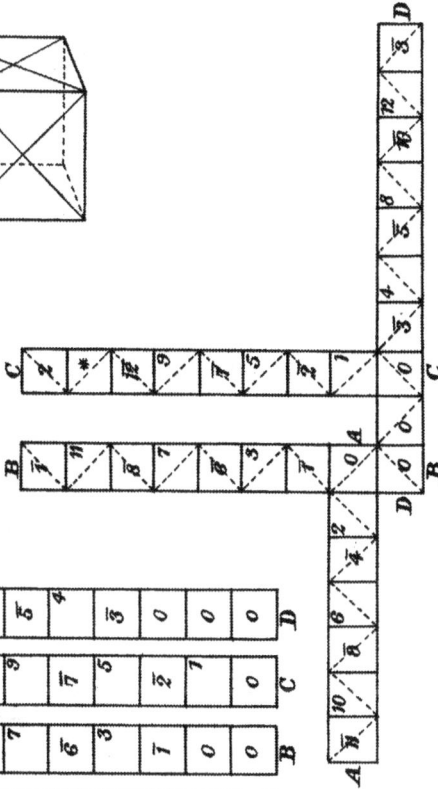

Fig. 10.

TYPICAL CUBE ADJUSTMENT.

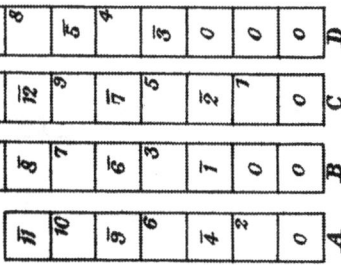

The curved strips in Fig. 8 are here set out in their true rectilineal order for final adjustment, as in Fig. 10.

Fig. 9.

E & F.N.Spon, London & New York

Plate 6

RHOMBIC DODECAHEDRON.

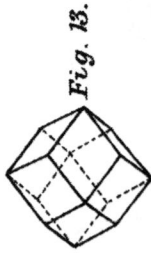

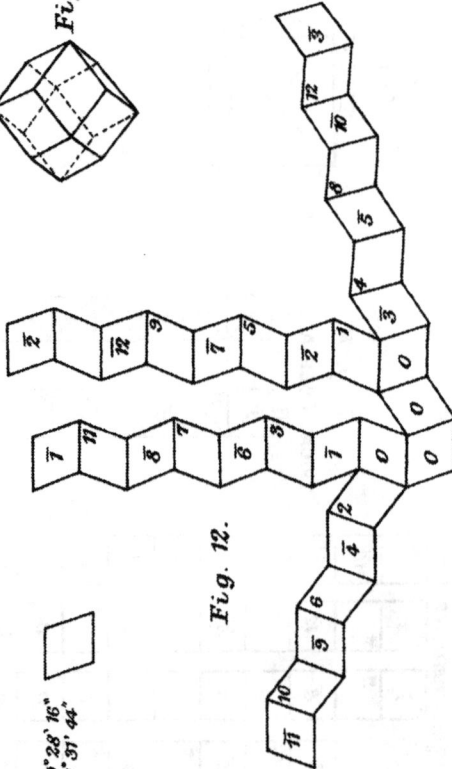

Fig. 13.

Fig. 12.

109° 28' 16"
70° 31' 44"

Plate 7

REGULAR OCTAHEDRON AND CUBE.

Fig. 14.

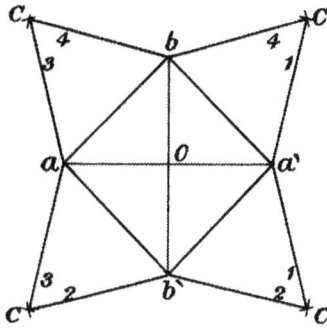

Fig. 15.

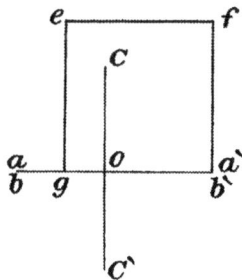

E & F.N.Spon, London & New York

THE FIRST SYSTEM.

Each system is defined in accordance with the relative *length* of its *axes,* and the angle of their intersections. In the first system the angles are *equal* in length, and cut in the centre at *right* angles. (For cube, see Fig. 6.)

Regular Octahedron.—A solid bounded by eight faces, which are equilateral triangles; inclination of faces at their edges equal to 109° 28′ 16″; edges, twelve; solid angles, six.*

It is proposed to elicit from the axes—

1. The planes or faces of the model;
2. The method of locating their sides; and then from the cube itself—
3. The mode of adjusting the faces to form a plaited model of the octahedron.

To find a face of the octahedron, although this is known from the above definition, it is proposed to obtain it direct from the axes, in order that the same method may be employed in the case of other forms which are more complicated, and where the faces are not known. Let the straight lines *a—a′, b—b′*, Fig. 14, be called the first and second axes of the system, and let *c—c′*, Fig. 15, be the third axis. For convenience' sake, let each of them measure 2·2 inches.

Assuming that every octahedron is composed of two four-sided pyramids, having one common base, join *a b a′ b′*, Fig. 14, by straight lines: the common base is equal to *a b a′ b′*, Fig. 14.

To find the *triangular faces from the axes:*—Let *c o*, Fig. 15, be equal to half the third axis, and let the lines *a a′, b b′* represent the first and second axes, and let it cut *c o* at right angles.

With radius equal to *a c*, Fig. 15, set off the arc *a c*, Fig. 14; and with radius *b c*, Fig. 15, set off the arc *b c*, Fig. 14. The two arcs shall cut in *c*, and the *equilateral triangle a b c* is the *face required.*

The three remaining faces are set off upon their common base in the same way by reference to the letters.

Here, then, we have four faces projected upon their common base, and forming, when brought together at their sides, one of the pyramids. Each

---

* Listing explains and generalises the so-called *theorem of Euler about polyhedra,* viz. that if S be the number of solid angles of a polyhedron, F the number of its faces, and E the number of its edges, then S + F = E + 2. See Prof. Tait on 'Listing's Topologie,' p. 46.

triangle is now distinguished by using numerals, which are so disposed that any two sides which are marked with the same number will correspond in length, and form an edge when united. By this arrangement there is obtained not only the actual magnitudes of the sides of the faces of the octahedron, but also there is given a name to the sides of every two adjoining faces which are to form an edge by their union. This method provides the limit to what *can* and what *cannot* be accomplished by the process of plaiting: it *can*, and does, furnish us with the true adjustment of the faces themselves in making a model; but it *cannot* locate the sides of those faces. This requires the use of arbitrary numbers, each side being represented by its own numbers; and when this is effected in the case of the regular octahedron, it answers for all the rest.

On contrasting this solid with the *cube*, we notice that two forms totally distinct from one another—the one bounded by *six* faces, the other by *eight*—are adjusted in precisely the same numerical order. This was solved by taking an *obtuse rhombohedron,** the plane angles of the faces of which were equal to 60° and 120°. When creased in their short diagonals, these faces became converted into twelve equilateral triangles, exceeding by four, be it observed, the number actually required to make the octahedron. It was found, however, that this extra number exactly corresponded with that which was required for the foundation and finishing planes of the model. From this peculiarity in its construction it results that *each pair of triangles must be counted as one plane* throughout. Thus constructed, the manipulation will be found to be exactly that of the cube, and every octahedron from the *regular* to the *doubly oblique* may be made up with the same facility.

To show that the *quasi rhombohedron* from which those speculations and results have been obtained, has no actual existence, see Figs. 78 and 79.

---

* Not the solid itself which with plane angles of this value is an impossible form, but the planes projected as if for any other obtuse rhombohedron, for the projection is the same in all.

Plate 8

REGULAR OCTAHEDRON.

Fig. 16.

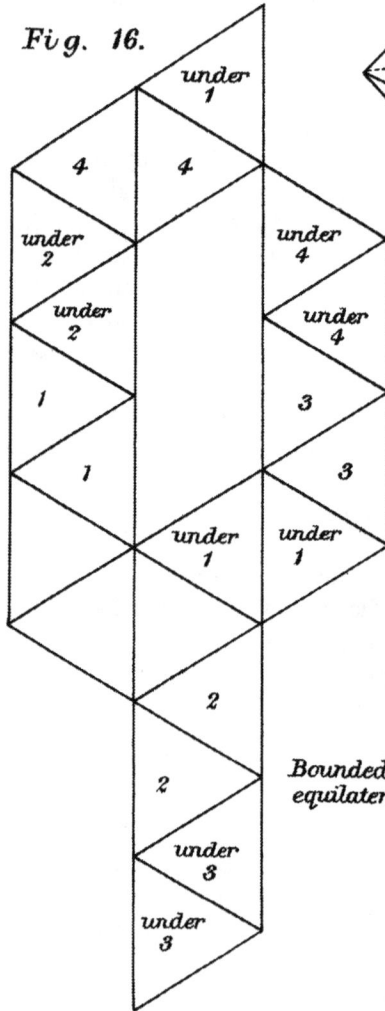

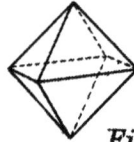

Fig. 17.

Bounded by 8 similar equilateral triangles.

E & F.N.Spon, London & New York

Plate 10

TYPICAL OCTAHEDRON.

Fig. 18 A.

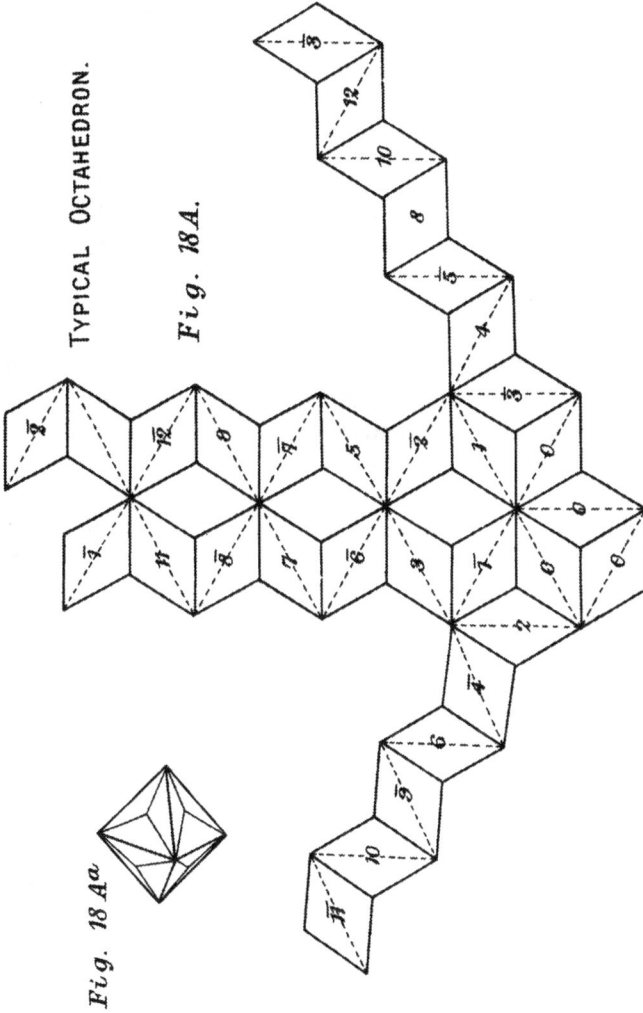

Fig. 18 A^a

E & F.N.Spon, London & New York

MODE OF USING THE KEY

On the following page a key is given for making all the octahedrons, and on the opposite side the faces or planes under which they are contained. To use this key, a single glance will suffice to show that each face (which is supposed to be cut in cardboard), when traced round on a piece of cartridge-paper exactly in the order prescribed by the key, will make in the aggregate the projected model of its own octahedron. The numbers on the card pattern are sometimes made on one side; if on both, the card is reversed, and the required numbers are then appended as shown in the triangles of dotted lines.

It will be noticed in the key that certain of the faces are *shaded*; this is done to indicate that in tracing the outlines of such faces the *card pattern* is to be *reversed;* for on carefully examining the two opposite pyramids set on a common base of any octahedron, it will be found that while the faces of the two pyramids are all exactly alike, those of the one pyramid are always the reverse of the faces of the other pyramid.

Typical Octahedron.—So called, because by its use pyramids of a given height may be raised on all the faces of the *regular octahedron*. In its present form the three isosceles triangles composing a face are equal to 120°; these when meeting lie in one plane, and the model is that of a regular octahedron, with each face composed of three pieces instead of one. On making the isosceles angles equal to 118° 57' 55", a low pyramid is obtained, and the form becomes that of the *octahedral fluor haloide*, and is called an *octahedral trigonal isositetrahedron*—that is, a crystal of twenty-four isosceles triangles, having an octahedral aspect. If the isosceles angles are diminished to 109° 28' 16", the well-known *rhombic dodecahedron* is the result.

The projection, Fig. 18A, consists of thirty-two similar and equal rhombs of 60° and 120°, adjusted as in the figure. Each rhomb is creased in the direction of the dotted lines, and is plaited by bringing the face numbered 1 over the opposite face numbered $\bar{1}$, and so on in numerical order until the model figure A *a* is obtained.

Tetrahedron.—The tetrahedron, a regular solid of geometry, is contained under four equilateral triangles, and therefore all its plane angles are equal to 60°. The faces incline to each other at the edges at an angle of 70° 31' 44".

The projection for making the plaited model of this solid is shown in the subjoined Fig. 19:—

The tetrahedron is a half form (hemihedral), that is, has half the number of faces of the regular octahedron, the alternate faces of which when enlarged to infinity cut one another, and produce a model of four instead of eight faces.

Sphenoids.—A sphenoid (from σφην, ηνὸς, *cuneus*, a wedge) is an irregular tetrahedron; it bears the same relation to *its allied* octahedron, of which it is the hemihedral form, as the tetrahedron does to the *regular* octahedron. The sphenoids resemble the tetrahedron, in having the same number of triangular faces; but the triangles are never equilateral. They are projected for modelling like those of the tetrahedron.

Tetragonal Sphenoid is bounded by four similar and equal isosceles triangles, and is derived from the *right square octahedron*. (Fig. 20.)

Right Rhombic Sphenoid.—Bounded by eight similar and equal scalene triangles—derived from *right rhombic octahedron*. (Fig. 21.)

Oblique Sphenoid.—Bounded by scalene triangles of two kinds—derived from *oblique rhombic octahedron*. (Fig. 22.)

Doubly Oblique Sphenoid.—Bounded by scalene triangles of four kinds— derived from *doubly oblique rhombic octahedron*. (Fig. 23.)

Tetrahedron (Typical).—For crystal purposes this form is best made by resolving each of its triangular faces into three *isosceles* triangles of 120°, which when plaited meet in the centre, and form one plane. When the isosceles angles are less than 120°, they become elevated from the centre, and rise in the form of a three-sided pyramid on each face. In *tetrahedral copperglance*, for instance, the angles are equal to 117° 2' 8", and the pyramid is low. In *dodecahedral garnet-blende* the isosceles angles are reduced to 112° 63' 7", the pyramids becoming raised in proportion. The limit to these modifications consists in making the isosceles angles equal to 90°; the resulting form is a *cube*. By using this method of making the tetrahedron, the relation between it and the cube, and the transition from the one form to the other, are well displayed. It is essentially a typical model representing something which is capable of being evolved on one and the same type, by the mere alteration of the plane angles of its isosceles faces. (Fig. 24)

Triangular Prism is contained under five planes—three lateral, which are squares or rectangles, and two terminal, which are equilateral triangles. From the fact of its being used by Newton to decompose white light into its seven primitive colours, this form is familiar to every one. (Fig. 25.)

To make the model, the planes are projected in the same order as in the cube.

Plate 9

Key to the adjustment of the faces of all the Octahedrons.

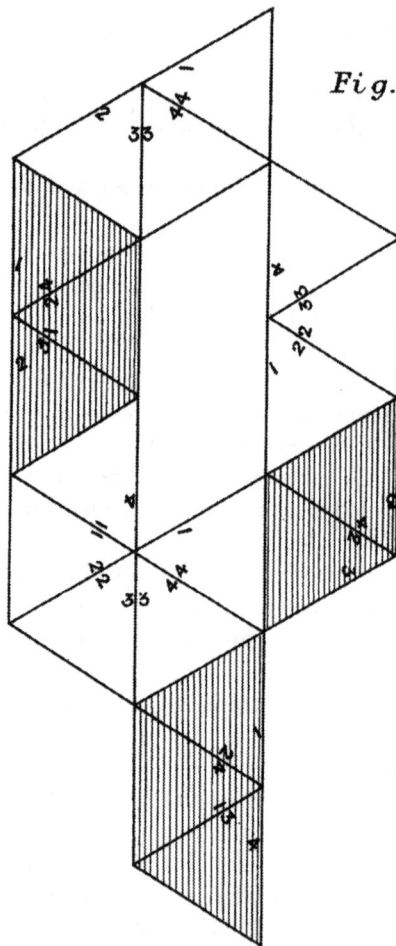

Fig. 18.

E & F.N.Spon, London & New York

Plate 13

FACES OF ALL THE OCTAHEDRONS OF AXIAL SYSTEMS. FOR KEY.

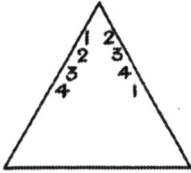

Req: Octahed: (1card)   Rt: squ: Octa: (1 card)   Rt: rhom: Octahe (1 card)

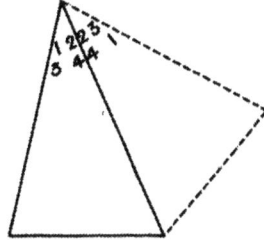

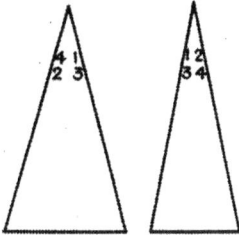

Rt: rect: Octahe: (2cards)   Oblique rhomb: Octahedron (2 cards)

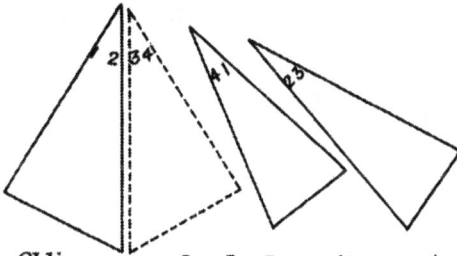

Oblique rect: Octahedron (3 cards)

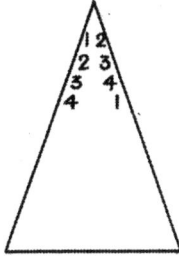

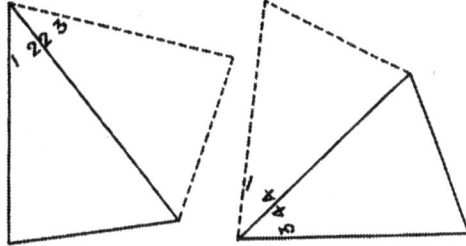

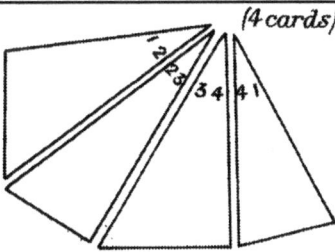

(4 cards)

Doubly ob: rh: oct: quasi rt. rect.   Doub: ob: rhomb: oct: (4 cards)

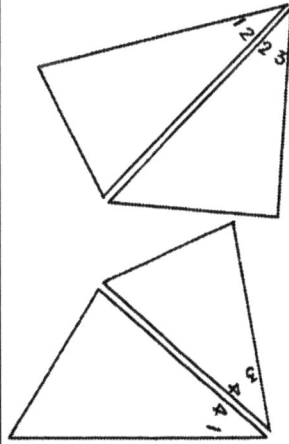

E & F.N.Spon, London & New York

Plate 11

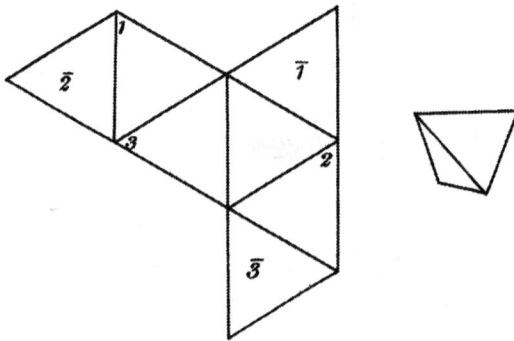

Fig. 19.

E & F.N.Spon, London & New York

Plate 12

Fig. 20.

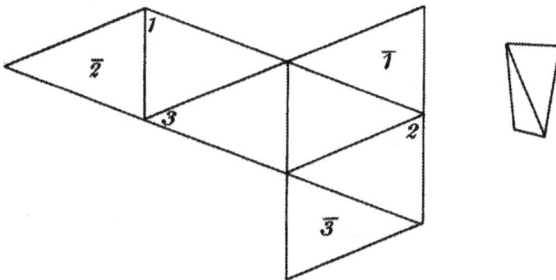

Fig. 21.

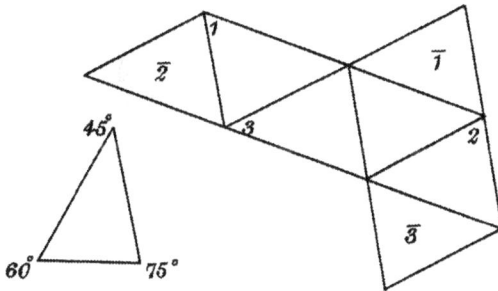

E & F.N.Spon, London & New York

Plate 56

Fig. 22.

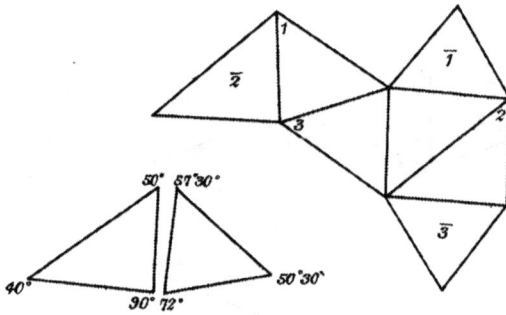

Fig. 23.

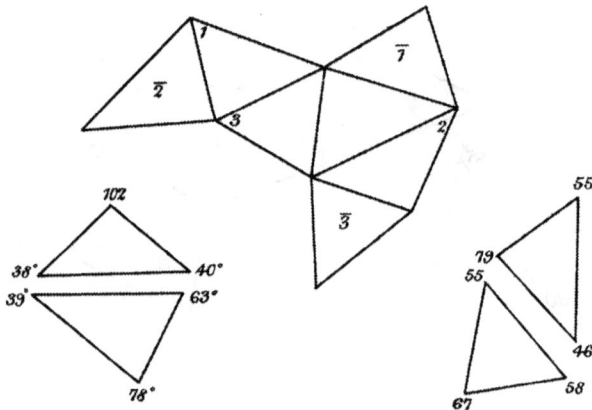

E & F.N.Spon, London & New York

Plate 17

Right Square Octahedron.

Fig. 29.

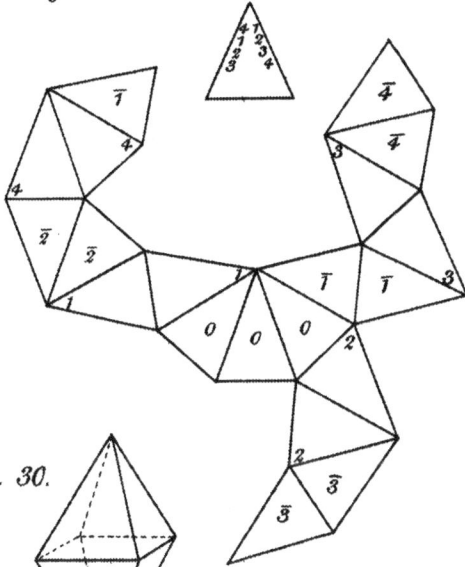

Fig. 30.

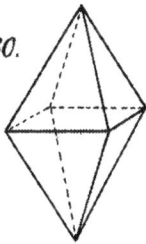

E & F.N.Spon, London & New York

Plate 16

Fig. 26.

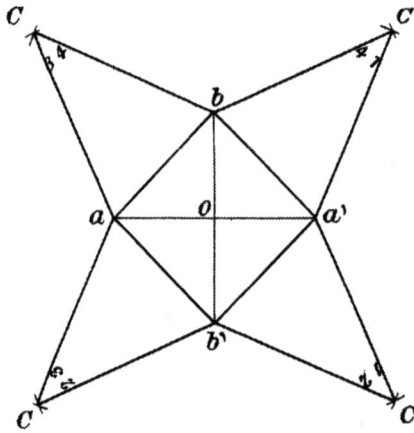

Fig. 27.

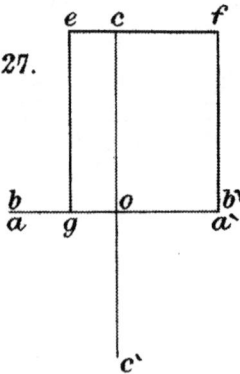

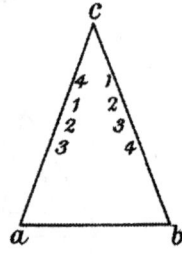

Fig. 28.

E & F.N.Spon, London & New York

Plate 15

Fig. 25.

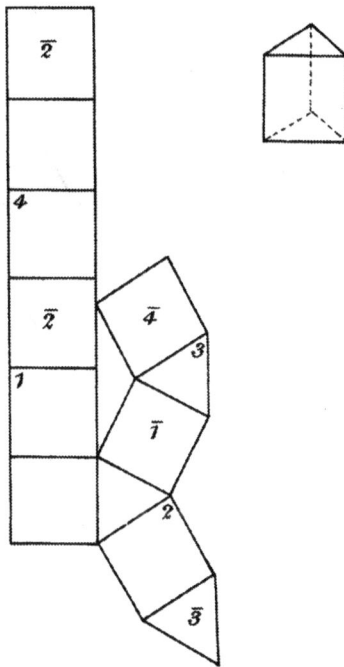

Plate 14

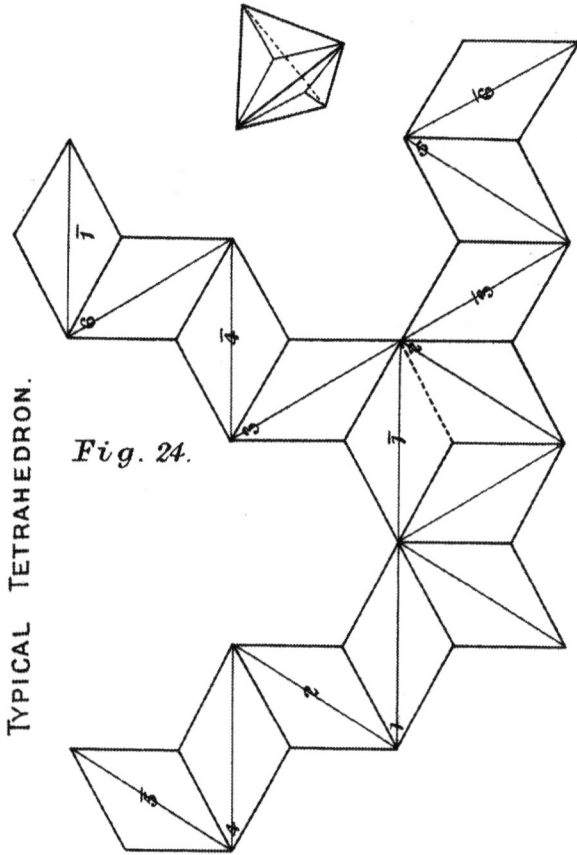

TYPICAL TETRAHEDRON.

Fig. 24.

E & F.N.Spon, London & New York

SQUARE PRISMATIC SYSTEM

has three axes, all at right angles, one longer or shorter than the other two. (In our examples the third axis is taken longer than the other two.)

> *Forms:—Right Square-based Octahedron.*
> *Right Square Prism.*

Right Square Octahedron.—To find the *base* (Fig. 26):—Draw a—a' and b—b' (the first and second axes), equal to 2·2 inches. Because they are equal and rectangular, the square formed by joining their extremities by four straight sides is the *base of the octahedron*, that is, the base common to the two isosceles pyramids composing the solid.

To find the *triangular face* (Fig. 27):—Draw c—c' (the third axis) equal to 3.3 inches, and let it be bisected in o, and let a—a' (the first axis) be perpendicular to $c\,o$. With radius $a\,c$ place one foot of compasses in a, Fig. 26, and describe the arc c; with same radius describe the arc $b\,c$. The triangle $a\,b\,c$ is a face of the octahedron. Because the faces are all equal, the three remaining triangles are found in the same way by using the same radius. To distinguish the sides of the faces the numerals are now added; and these are repeated on the cardboard pattern of the face, Fig. 30. This pattern being the same for all the outlines, would imply indifference as to position; but reference to the key, Fig. 18, shows that the plan of each side of every plane is fixed and definite.

Right Square Prism (Fig. 32).—Bounded by six planes, four of which are rectangular, and two square.

To find the *square planes* from the axes:—Draw the first and second axes a—a' and b—b', Fig. 26, intersecting in o at right angles. Join the ends of the axes by straight lines; then $a\,b\,a'\,b'$ is a *square plane* of the prism.

To find a *lateral plane* of the prism by measurement from the axes:—Draw $c\,o$, half the third axis, and a—a', the first axis, at right angles to one another. With compasses take $a\,b$, a side of the square base, Fig. 26, and measure this distance along the line $a'\,g$, Fig. 27. The parallelogram $a'\,g\,e\,f$ is the *lateral plane*. To make the model, these two faces are adjusted like those of the cube, as in the following Fig. 31; Fig. 33, terminal plane; Fig. 34, lateral plane.

RHOMBIC SYSTEM

has three axes of unequal length, at right angles to each other.

>   Forms :—*Right Rhombic-based Octahedron.*
>   *Right Rhombic Prism.*
>   *Right Rectangular-based Octahedron.*
>   *Right Rectangular Prism.*

Right Rhombic Octahedron.—To find the *base*:—Let the first and second axes a—a' and b—b', Fig. 33, cut at right angles in o. Join the extremities by straight lines; the rhomb a b a' b' is the *base* of the octahedron.

To find a *triangular face*:—Draw the half of the third axis, c o, Fig. 36, equal to 1·8 inch, and the second axis b—b', equal to 2·2 inches. Make the first axis a—a' equal to 1·6 inch, and let it be measured along the second axis b—b'.

Plate 25

RIGHT RECTANGULAR PRISM DERIVED FROM RIGHT RHOMBIC PR.

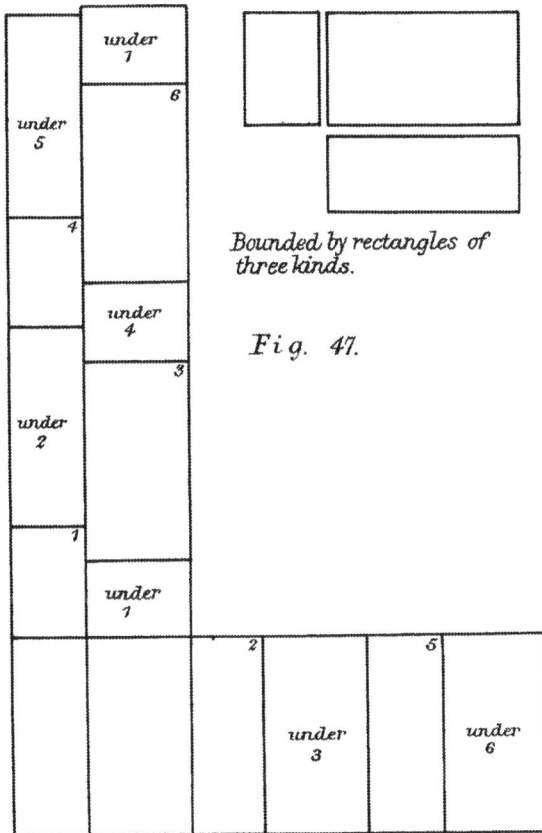

under
7

6

under
5

4

under
4

under
2

3

1

under
1

2

5

under
3

under
6

Bounded by rectangles of
three kinds.

Fig. 47.

Plate 24

Right Rectangular Octahedron Derived From Right Rhombic Octahedron.

Fig. 45.

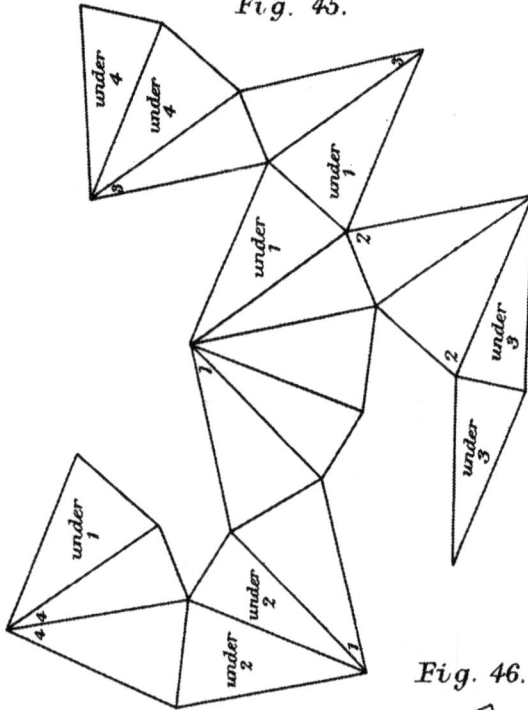

Fig. 46.

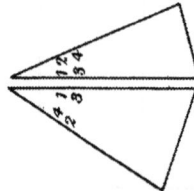

Bounded by isosceles triangles of two kinds.

E & F.N.Spon, London & New York

Plate 23

RIGHT RECTANGULAR OCTAHEDRON AND PRISM.

Fig. 43.

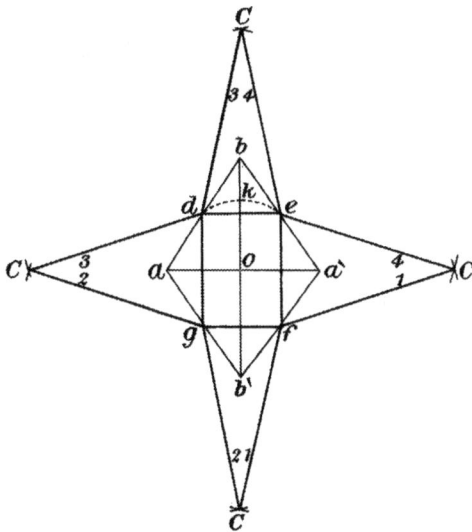

Fig. 44.

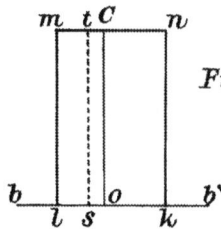

E & F.N.Spon, London & New York

Plate 22

RIGHT RHOMBIC PRISM
THIRD AXIS SHORTER THAN THE OTHER TWO.

Micros: Crystal Uric acid.

Fig. 40.

Fig. 42.

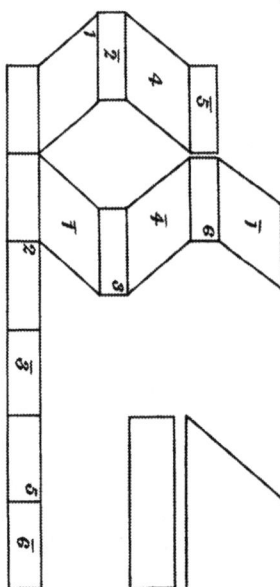

Fig. 41.

E & F.N.Spon, London & New York

Plate 21

RIGHT RHOMBIC PRISM.

Fig. 39.

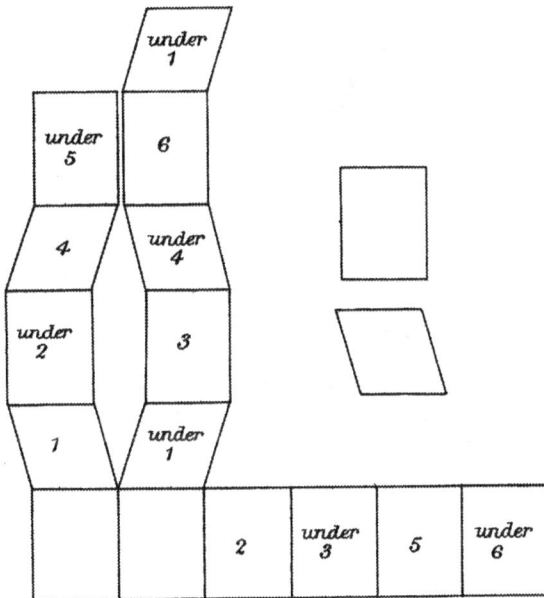

E & F.N.Spon, London & New York

Plate 20

RIGHT RHOMBIC OCTAHEDRON.

Fig. 38.

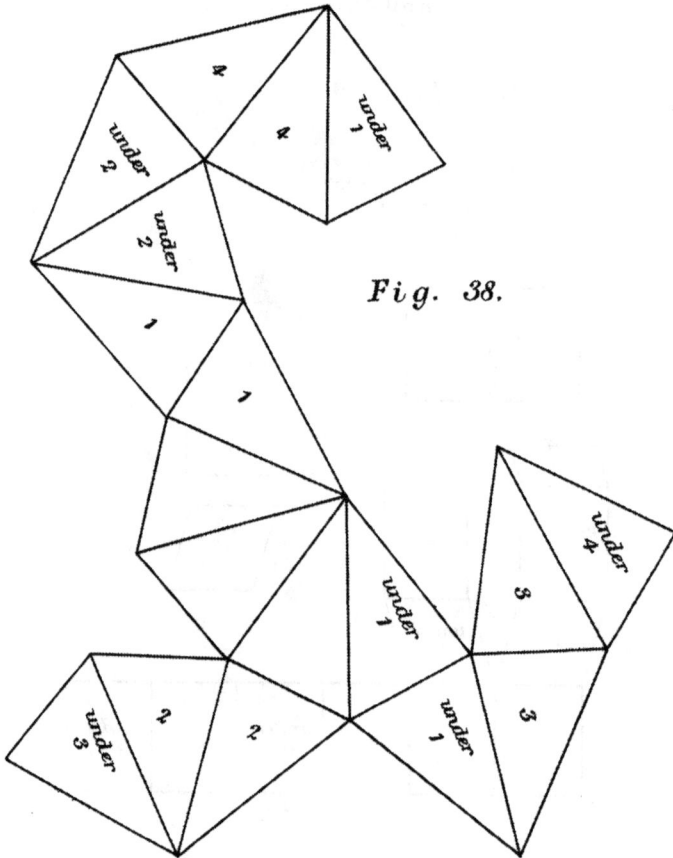

E & F.N.Spon, London & New York

Plate 19

RIGHT RHOMBIC OCTAHEDRON AND PRISM.

Fig. 35.

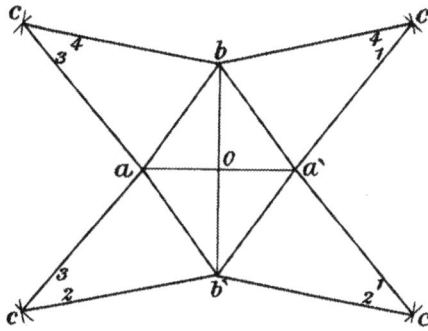

Fig. 37.   *Fig. 36.*

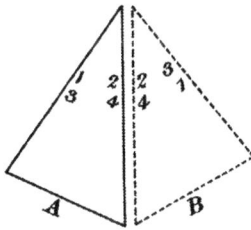   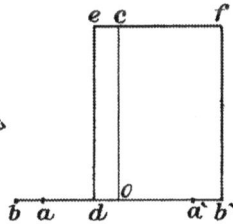

E & F.N.Spon, London & New York

Plate 18

RIGHT SQUARE PRISM.

Fig. 31.   *Fig. 32.*

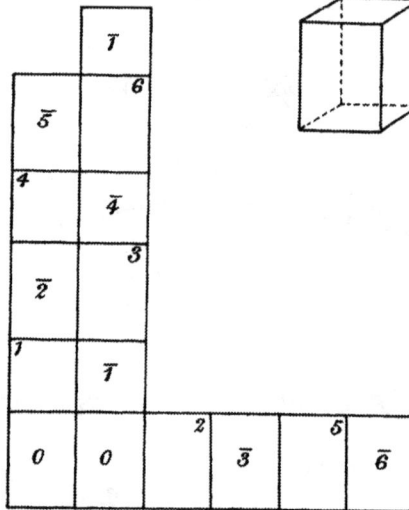

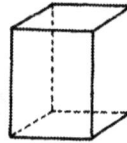

Fig. 33. 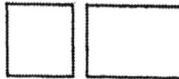 *Fig. 34.*

E & F.N.Spon, London & New York

With compasses and with radius $c\, a$, Fig. 36, describe the arc $a\, c$, Fig. 35, and with radius $b\, c$, Fig. 36, describe the arc $b\, c$, Fig. 35; the two arcs shall cut in c, and the *scalene* triangle $a\, b\, c$ is a face of the octahedron. Similarly, let the remaining triangles be set off from the sides of the rhombic base by reference to the letters in Fig. 36, and let each triangle be supplied with numerals exactly in the same order as in Fig. 26.

To adjust the faces by the key, Fig. 18, so as to obtain the projection of the model:—Cut out of cardboard the triangle $c\, a'\, b'$, Fig. 35. This constitutes the pattern round which the tracings are made of every triangle in the model.

On the *obverse* face of the pattern write the numbers $\left\{ \begin{array}{l} 1,\ 2 \\ 3,\ 4 \end{array} \right.$; and on the

reverse face, shown by the dotted lines, the numbers $\left\{ \begin{array}{l} 2,\ 3 \\ 4,\ 1 \end{array} \right\}$. (Fig. 37.)

Now let outlines of the pattern be traced on a sheet of cartridge-paper, so that the numbers on the pattern and on the key shall correspond, being careful that *where the planes are shaded the pattern is reversed* before making its outlines.

Finally, having obtained the projection, Fig. 38, let the order in which the faces are to be plaited into the model be indicated by the numbers as shown in Fig. 16.

To find the six planes of the *right rhombic prism:*—These consist of two terminal faces which are *rhombs*, and four lateral planes which are *rectangles*.

To find the *rhombic or terminal faces;*—Let the first and second axes a—a' and b—b' intersect in o, Fig. 35, at right angles. Join their extremities by straight lines; the rhomb thus made is the *terminal face* required.

To find the lateral planes of the right rhombic prism:—On the line $b'\, b$, Fig. 36, measure off $b'\, d$ equal to $a\, b$, Fig. 35. Make $d\, e$, Fig. 36, equal to $c\, o$, and describe the parallelogram $b'\, d\, e\, f$, which is the *required lateral plane.*

To adjust the planes on a plane surface for the purpose of plaiting them into the model, see Fig. 39.

Right Rectangular Octahedron—derived from the right rhombic octahedron—bounded by eight isosceles triangles of two kinds.

To find the *base:*—Bisect the sides of the rhomb *a b a' b'*, Fig. 43. Join the bisecting points by straight lines; the rectangle *d e f g* is the *base* of the octahedron.

To find the *faces* :—With radius *o d*, Fig. 43, draw the arc (dotted lines) *d e*, it shall cut *b b'* (the second axis) in K. Measure *o* K along *b b'*, Fig. 44; then *c* K is equal to any side of an isosceles face, and the triangles *c d e* and *c d g* are the two faces of the octahedron.

Right Rectangular Prism is contained under six faces, which are rect-angles—two terminal and four lateral.

To find the two *terminal planes:*—These are equal to the base of the octa-hedron *d e f g.*

To find the two *broad lateral planes:*—Along the line *b b'*, Fig. 44, measure K *l* equal to *d g*, Fig. 43, and describe the parallelogram K *l m n,* which is equal to the required plane.

To find the *narrow lateral planes:*—On the line *b b'*, Fig. 44, measure K *s* equal to *d e*, Fig. 43. The parallelogram K *s t n* is equal to the lateral plane required.

OBLIQUE PRISMATIC SYSTEM.

Three axes, all unequal in length: two, *a—a'* and *b—b'*, at right angles; the third, *c—c'*, oblique to one and perpendicular to the other.

Forms:—*Oblique Rhombic-based Octahedron.*
Oblique Rhombic Prism.
Oblique Rectangular-based Octahedron.
Oblique Rectangular Prism.

Oblique Rhombic-based Octahedron.—Contained under eight *scalene* tri-angles of two kinds.

To find the *base:*—Draw the first axis (*a—a'*) equal to 1.6 inch, and the second axis (*b—b'*) equal to 2.2 inches, at right angles to one another. Let the extremities be joined by straight lines. The rhomb *a b a' b'* is the base.

To find the two *triangular faces* (Fig. 49):—Draw *c o* (half the third axis) equal to 1·8 inch, *a—a'* (first axis) equal to 1·6 inch, at right angles to one another. Let *b—b'* (the second axis) cut *c o* in *o* obliquely, so that *c o b'* shall be equal to 120°.

Plate 33

OBLIQUE RECTANGULAR PRISM.

Fig. 63.

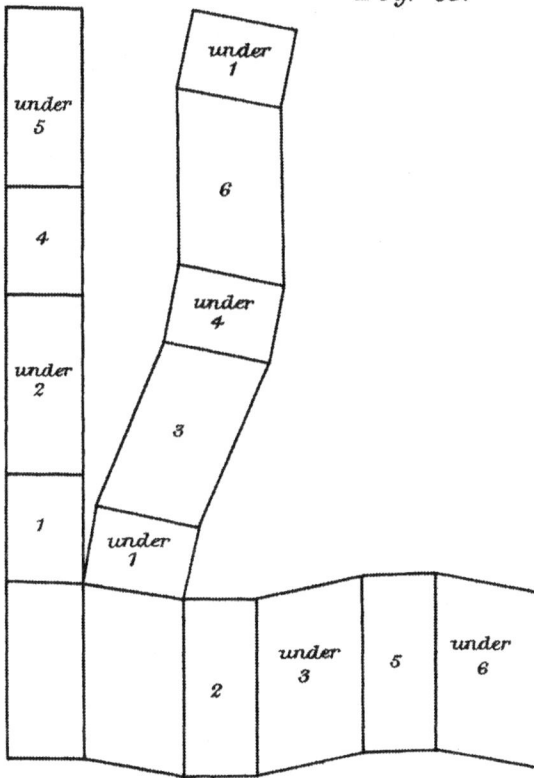

E & F.N.Spon, London & New York

Plate 32

OBLIQUE RECTANGULAR OCTAHEDRON.

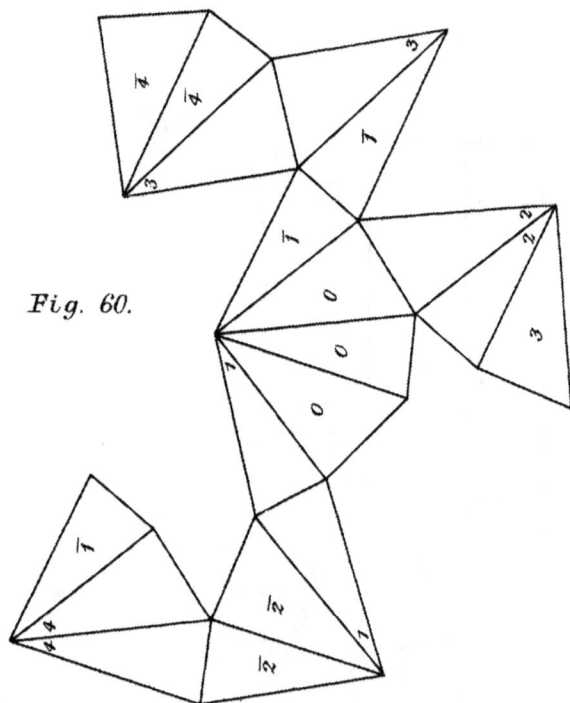

Fig. 60.

Fig. 62.

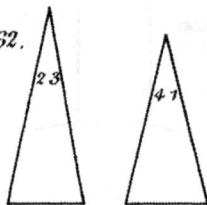

Fig. 61.

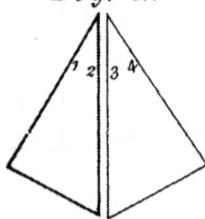

E & F.N.Spon, London & New York

Plate 31

OBLIQUE RECTANGULAR OCTAHEDRON AND PRISM.

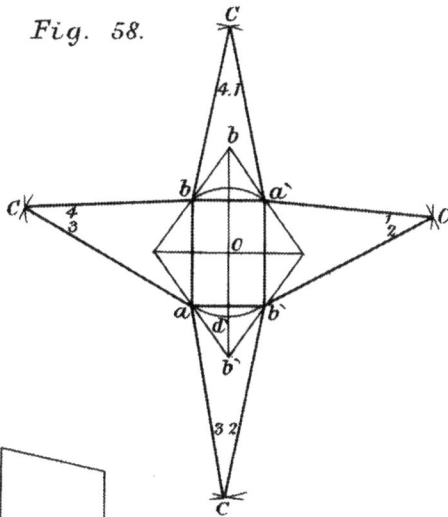

Fig. 58.

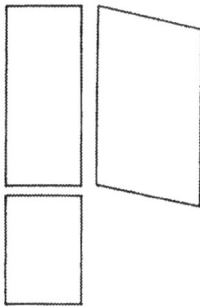

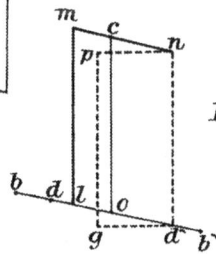

Fig. 59.

E & F.N.Spon, London & New York

Plate 27

OBLIQUE RHOMBIC OCTAHEDRON. *(coḃ = 120°)*

Fig. 50.

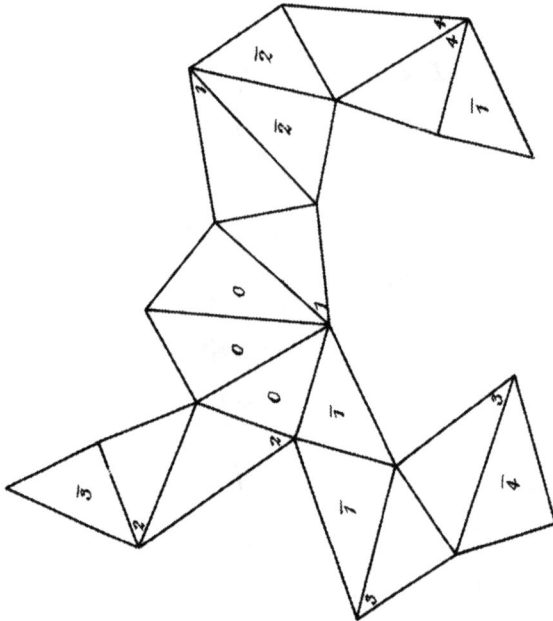

Fig. 51

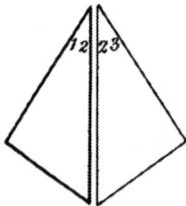

Fig. 52.

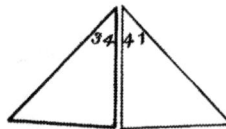

E & F.N.Spon, London & New York

Plate 28

OBLIQUE RHOMBIC OCTAHEDRON. *(cob'= 100°)*

Fig. 53.

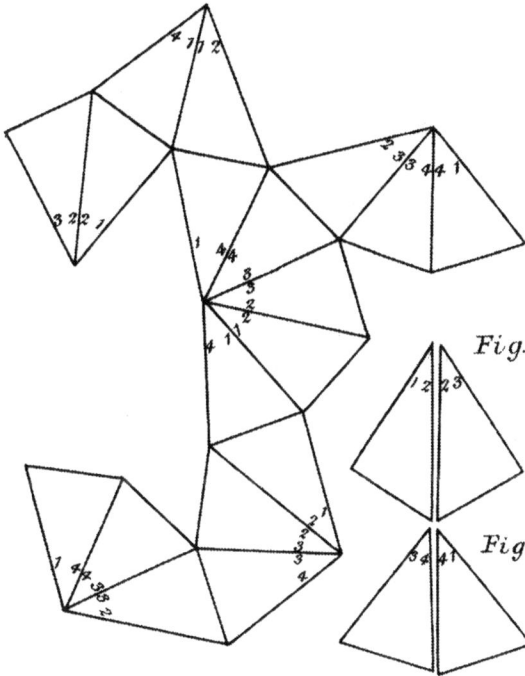

Fig. 54.

Fig. 55.

E & F.N.Spon, London & New York

Plate 30

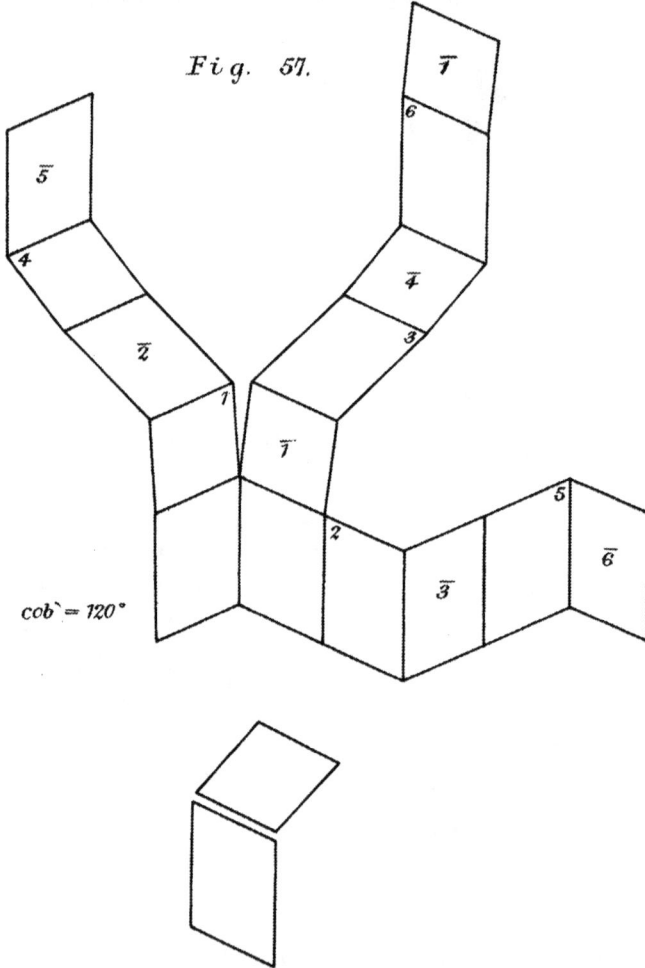

OBLIQUE RHOMBIC PRISM (VAR. 2)
(OBLIQUITY IN SHORT DIAG: OF TERMINAL PLANE.)

$Fig.\ 57.$

$cob` = 120°$

E & F.N.Spon, London & New York

Plate 29

OBLIQUE RHOMBIC PRISM VAR. I.(OBLIQUITY IN LONG DIAGONAL)
PLANES SAME IN BOTH VARIETIES.

Fig. 56.

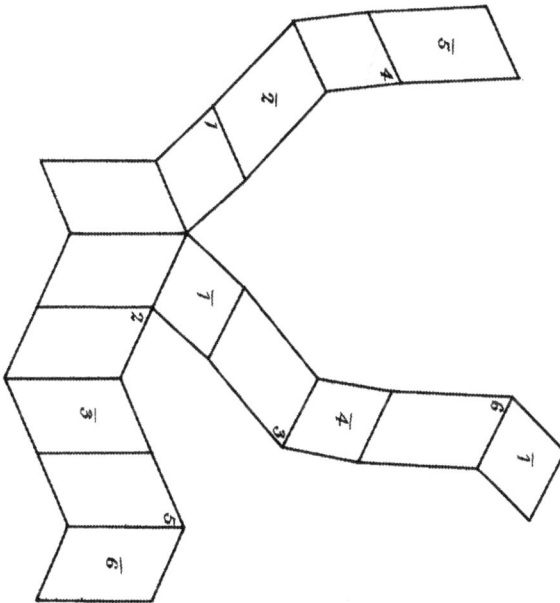

E & F.N.Spon, London & New York

Plate 26

OBLIQUE RHOMBIC OCTAHEDRON AND PRISMS.

Fig. 48.

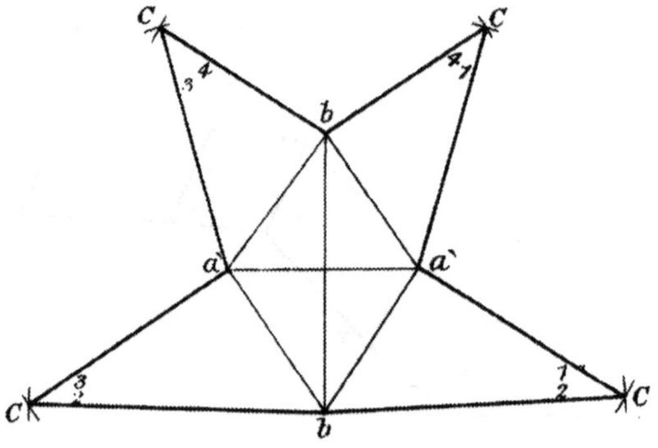

Fig. 49.

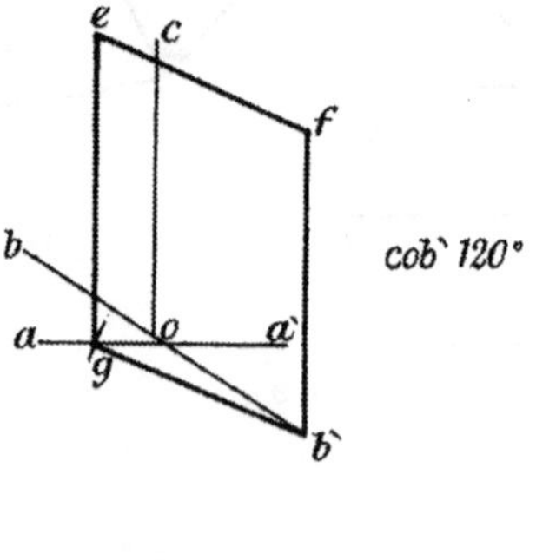

cob` 120°

E & F.N.Spon, London & New York

Take compasses, and with radius *a c* describe the arc *a c,* Fig. 48, and with radius *b c,* Fig. 49, describe the arc *b c,* Fig. 48. The two arcs shall cut in *c,* and the triangle *a b c* is a face of the octahedron. Similarly, let the other faces be set off upon the sides of the base, and append the numbers. Of the four faces thus formed, the two upper ones are obverse and reverse to one another, and the lower ones are the same. Hence the form is contained under faces of *two* kinds only.

Oblique Rhombic Prism.—Bounded by two terminal planes which are similar rhombs, and four lateral planes which are similar rhomboids.

To find the *terminal faces:*—Let the rhomb *a b a' b',* Fig. 48, be the terminal face.

To find the *lateral planes* (Fig. 49):—Draw *a—a'* (the first axis) at right angles to *c—o* (the third axis). Take *a b* any side of base, Fig. 48, and with one foot of compasses in *b',* Fig. 49, draw the arc *b g:* it shall cut *a—a'* in *g.* Upon *b' g* describe the parallelogram *b' g e f.* The required *lateral plane* is the rhomboid *b' g e f.*

Of this prism there are two varieties : in the first the obliquity takes place along the *greater* diagonal of the terminal rhombs; in the second variety the obliquity occurs in the *shorter* diagonal of the terminal planes. Both are obtained by altering the adjustment of the *same* planes.

Oblique Rectangular Octahedron.—Derived from oblique rhombic octahedron. Bounded by *isosceles* triangles of two kinds and *scalene* triangles of one kind.

To find the *base:*—Bisect the side of the rhombic octahedron from which it is derived; join the bisecting points by straight lines—the rectangle *a b a' b'* is the *base* of the octahedron.

To find the *isosceles* triangles (Fig. 59):—Let *c o* (half the long axis) = 1·8 inch, cut *b—b* (the second axis) equal to 2·2 inches at an oblique angle, so that *c o b'* shall be equal to 120°.

With compasses take the distance *o a',* Fig. 58, and with radius *o a'* describe the arc *a' b:* it shall cut *b b'* in *d;* with same radius describe the arc *a b':* it shall cut *b b'* in *d'.* Measure *d d',* Fig. 1, along the line *b b',* Fig. 2.

To find the *shorter triangular face:*—Take the distance *d c,* Fig. 59, and with one foot of compasses in b, Fig. 58, describe the arc *b c;* with same radius describe the arc *a' c;* they shall cut in *c.* The *shorter triangular face* is *a' b c.*

To find the *longer triangular face:*—Take the distance *d' c*, Fig. 59, draw the arc *a c*, Fig. 58, and with same radius draw the arc *b' c*; they shall cut in *c*; then *a b' c* is the *longer triangle*.

To find the *scalene triangle a b c:*—Upon the side *a b* of the rectangular base, Fig. 58, and with radius equal to side *a c* of the longer triangle, set one foot of compasses in *a*, and draw the arc *a c*; and with radius equal to side *b c* of the shorter triangle, place one foot of the compasses in *b* and draw the arc *b c*; the two arcs shall cut in *c*, and *a b c* is the *scalene triangle*. Append the numbers to each face.

Oblique Rectangular Prism.—Bounded by rhomboid lateral planes of one kind and rectangular planes of two kinds.

To find the *lateral rhomboid face:*—With radius *a b*, Fig. 58, set one foot of compasses in *d'*, Fig. 59, and measure *d' l* along the line *b b'*. Upon *d' l* describe the parallelogram *d' l m n;* then *d' l m n* is the *rhomboid face*.

To find the *lateral rectangular face:*—With radius *a' b*, Fig. 58, set one foot of compasses in *d'*, Fig. 59, and measure *d' g* along the line at right angles to *n d'*. Upon *d' g* describe the *lateral rectangular face d' g p n*, which was required.

To find the *terminal plane:*—Let *a b a' b'*, Fig. 58, be the *terminal plane*.

DOUBLY OBLIQUE PRISMATIC SYSTEM* (TRICLINIC-ANORTHIC).

Axes all unequal in length, and all oblique to each other.

Forms:—Doubly Oblique Rhombic Octahedron.
Doubly Oblique Rhombic Prism.

The Octahedron.—To find the *base:*—Let the short axis *a—a'*, equal to 1·6 inch, cut the second axis *b—b'* at an oblique angle, so that the angle *a o b* shall be equal to 100°. Join the extremities by straight lines ; then *a b a' b'* is a rhomboid, the common *base* of the two pyramids of the octahedron.

---

* "From their great apparent irregularity are exceedingly difficult to study and understand."—FOWNES.

Plate 34

DOUBLY OBLIQUE RHOMBIC OCTAHEDRON AND PRISM.

Fig. 64.

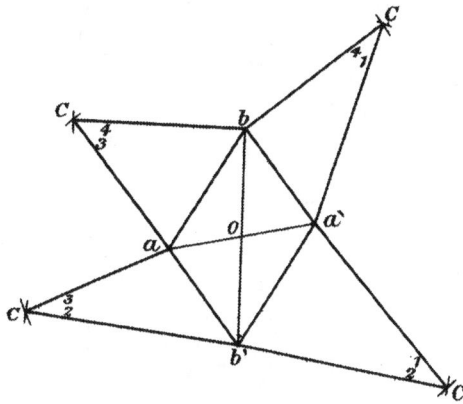

Fig. 65.

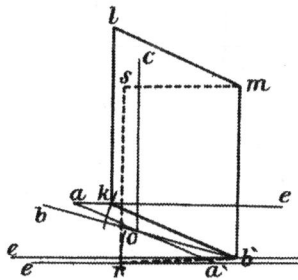

Plate 35

DOUBLY OBLIQUE RHOMBIC OCTAHEDRON.

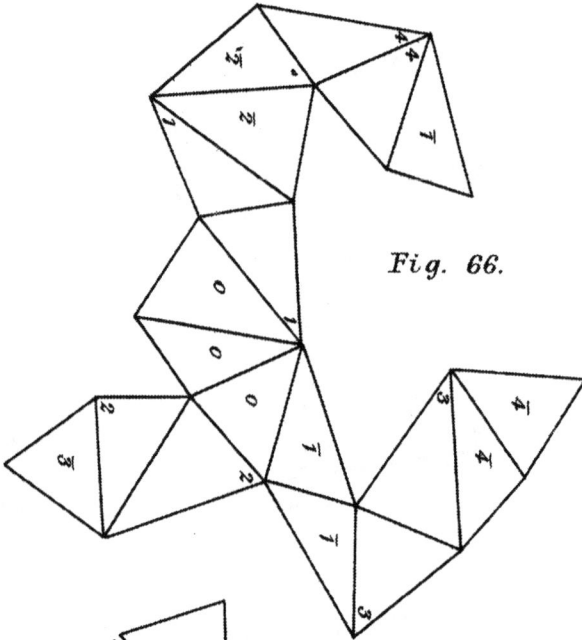

Fig. 66.

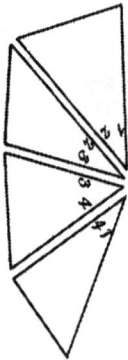

Fig. 67.

THE FOUR FACES.

E & F.N.Spon, London & New York

Plate 36

DOUBLY OBLIQUE RHOMBIC PRISM.

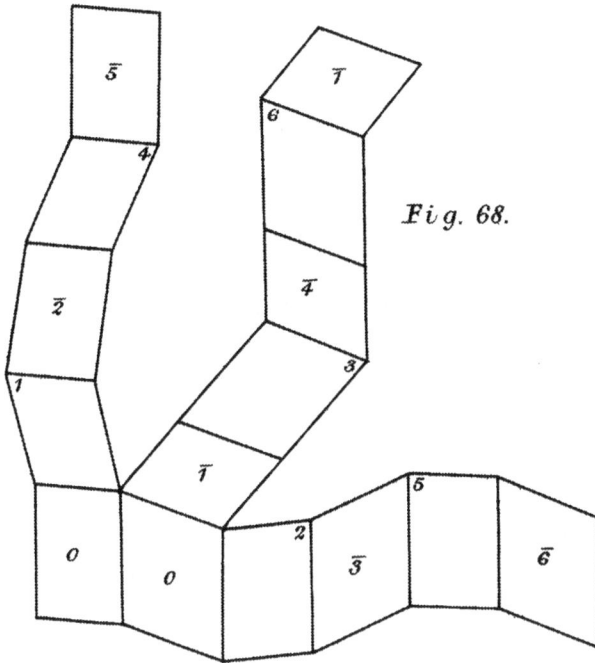

Fig. 68.

TERMIN. PLANE.

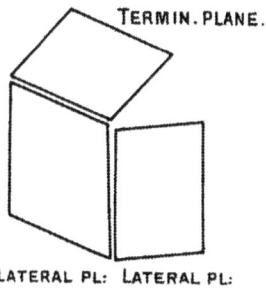

LATERAL PL: LATERAL PL:

E & F.N.Spon, London & New York

Plate 37

DOUBLY OBLIQUE DERIVED OCTAHEDRON AND PRISM.

Fig. 69.

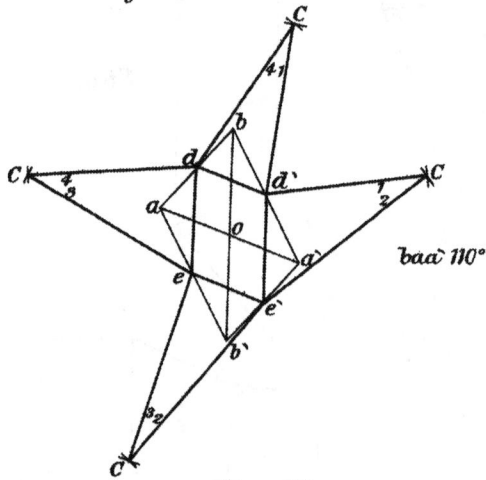

baa' 110°

Fig. 70.

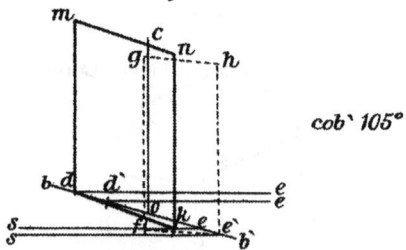

cob' 105°

E & F.N.Spon, London & New York

Plate 38

DOUBLY OBLIQUE OCTAHEDRON.
(QUASI RECTANGULAR)

Fig. 72.

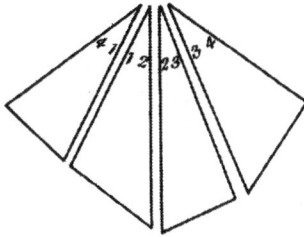

Fig. 71.

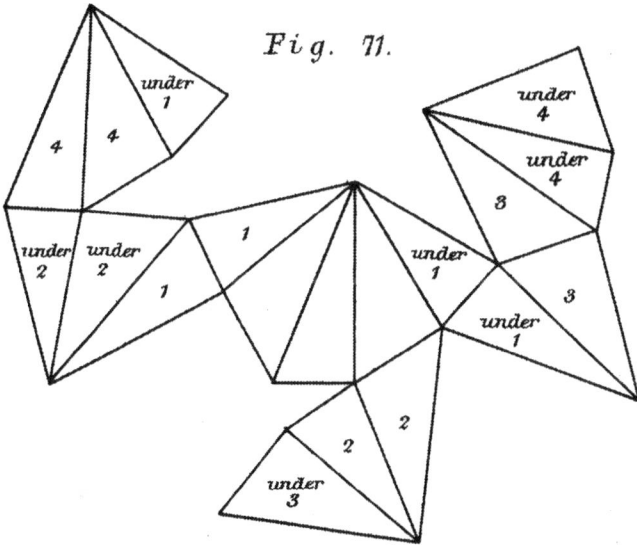

E & F.N.Spon, London & New York

Plate 39

DOUBLY OBLIQUE PRISM (QUASI RECTANG:)

Fig. 73.

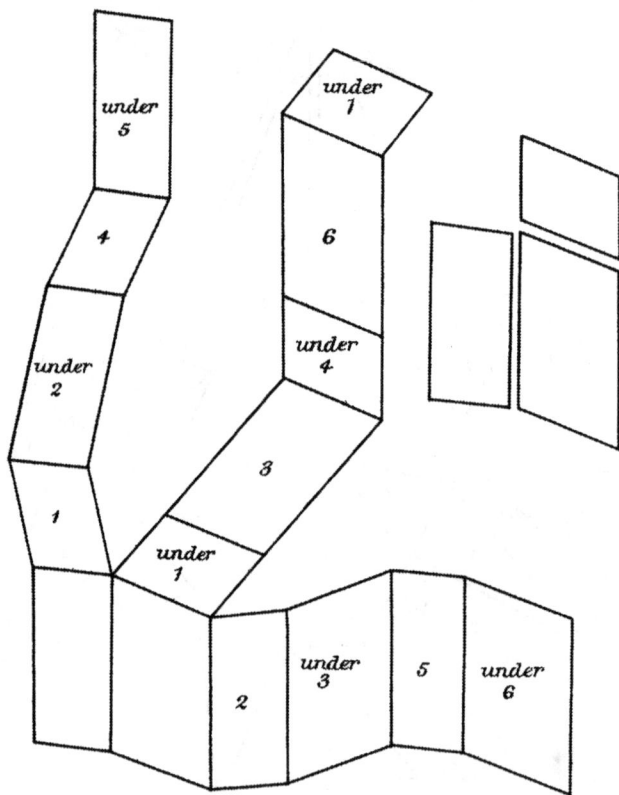

E & F.N.Spon, London & New York

To find the *triangular faces:*—Let *c o*, Fig. 65, be half the third axis, equal to 1·8 inch; let it be intersected in *o* by the short axis *a*—*a'*, equal to 1·6 inch, and at such an obliquity that the angle *c o a'* shall be equal to 110°. In like manner let the second axis *b*—*b'*, equal to 2·2 inches, intersect, so that the angle *c o b'* is equal to 104°.

With radius *c a* place one foot of compasses in *a*, Fig. 64, and draw the arc *a c*. Take *b a*, Fig. 65, and draw the arc *b c*, Fig. 64; the two arcs shall cut in *c*, and the *scalene triangle a b c* is one of the faces. To find a second face *a' b c*, proceed as before, the letters in both figures coinciding. The four *scalene triangles* are *a b a*, *a' b c*, *a' b' c*, and *a b' c*. Append the numbers for using the key.

The Doubly Oblique Rhombic Prism.—To find the *terminal planes:*—The base *a b a' b'*, Fig. 64, is one of the two terminal planes. Draw the lines *e a*, *e a'*, *e b'*, Fig. 65, at right angles to *c o'*, indicating the relative elevations of the two ends of the first axis *a a'*, and of the lowest end *b'* of axis *b b'*.

To find the *lateral face b' k l m* (Fig. 65):—With radius *a b*, Fig. 64, and foot of compasses in *b*, Fig. 65, describe an arc; it shall cut the line *e a* at *k*. Upon *b' k*, describe the parallelogram *b' k l m*, which is the rhomboid required.

To find the *lateral face b' r s m*, dotted lines (Fig. 65):—With radius *a' b*, Fig. 64, and one foot of compasses in *b'*, Fig. 65, describe an arc; it shall cut *b' r* in *r*. Upon *b' r* set the parallelogram *b' r s m*, which is the second lateral rhomboid of the prism.

A secondary octahedron and prism belong to this system. They take the place of the *rectangular* forms in the *oblique* rhombic system.

To find the *base* of the secondary octahedron (Fig. 69):—Let *a b a' b'* be the base of the primary form; bisect the sides, and join the bisecting points *d d' e' e* by straight lines; the rhomboid *e d' d e'* is the *base* required.

To measure the distances from *c* of the four angles of this rhomboid along the second axis *b* (Fig. 70):—With one foot of compasses in *o*, Fig. 1, take the distance *o d*, and measure off this distance *o d* along the second axis, Fig. 2. Similarly with radius *o d'*, Fig. 69, describe the arc *o d'* on the axis *b b'*, Fig. 70. In like manner set off *o e* and *o e'* on *b b'*, Fig. 70.

The four triangular faces can now be obtained:—With radius *d c*, Fig. 70, set one foot of compasses in *d*, Fig. 69, and describe the arc *d' c*. Similarly with radius *d' c*, Fig. 70, describe the arc *d c*, Fig. 69 ; the two arcs shall cut in *c*, and the *triangle d d' c* is one of the four faces. Let the sides of the remaining triangles be projected in the same way. The four *scalene* faces thus found are *c d d'*, *c e' d'*, *c e' e*, and *c e d*. They are adjusted to make the model (see Fig. 71).

DOUBLY OBLIQUE PRISM.

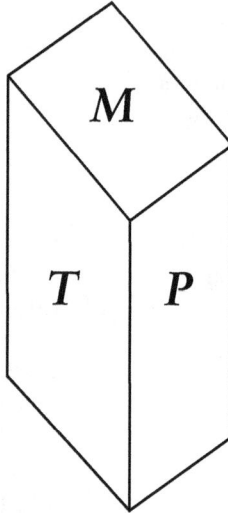

	Edges.			Plane Angles.	
		° ′			° ′ ″
Felspar	M on T = 120 35		·· in P =	123 29 0	
	P on M = 90 0		·· in T =	104 21 0	
	T on P = 67 15		·· in M =	63 18 30	
Kyanite	M on T = 106 15		·· in P =	105 56 0	
	P on M = 100 50		·· in T =	100 20 30	
	T on P = 93 15		·· in M =	90 15 0	
Diaspore	M on T = 64 54		·· in P =	58 27 0	
	P on M = 108 30		·· in T =	116 49 30	
	T on P = 101 20		·· in M =	112 40 30	

To find the *faces of the prism*:—Let *d e*, *d' e'* at right angles to *c o*, Fig. 70, be the lines of elevation of the angles of the rectangular base on the upper

Plate 40

Fig. 74.

$\bar{1}$

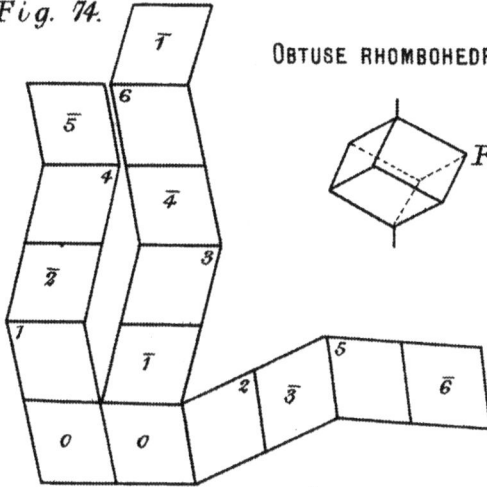

OBTUSE RHOMBOHEDRON.

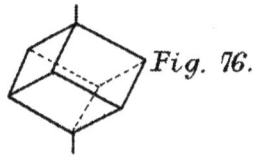

6

$\bar{5}$

Fig. 76.

4

$\bar{4}$

$\bar{2}$

3

1

$\bar{1}$

2 $\bar{3}$ 5 $\bar{6}$

0 0

ACUTE RHOMBOHEDRON.

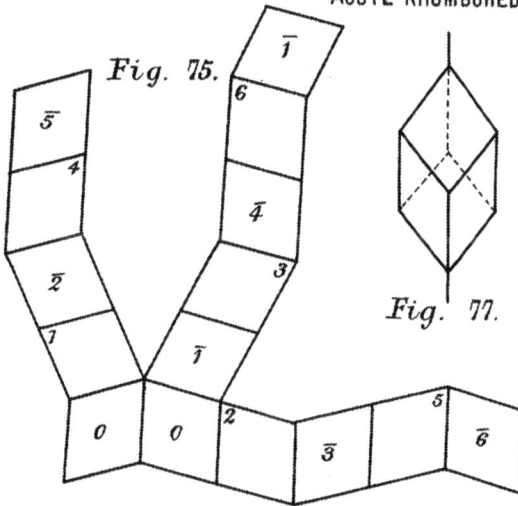

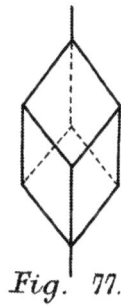

$\bar{1}$

Fig. 75.

6

$\bar{5}$

4

$\bar{4}$

$\bar{2}$

3

1

$\bar{1}$

Fig. 77.

0 0 2 $\bar{3}$ 5 $\bar{6}$

E & F.N.Spon, London & New York

Plate 41

A RHOMBIC PLANE OF 120 AND 60 THE LIMIT
TO THE OBTUSE RHOMBOHEDRONS.

Fig. 78.

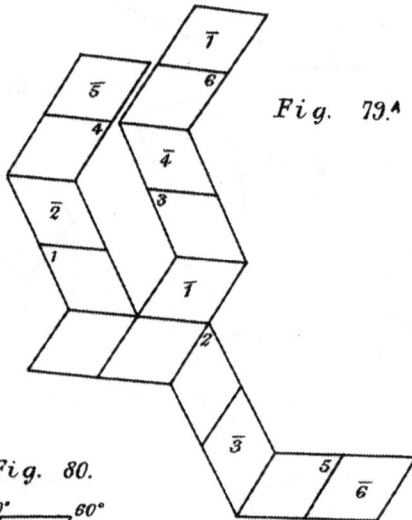

Fig. 79.ᴬ

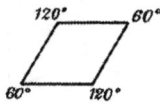

Fig. 80.

120°   60°

60°   120°

E & F.N.Spon, London & New York

Plate 42

QUASI OBTUSE RHOMBOHEDRON
PLANE ANGLES OF FACES EQUAL TO 60° AND 120°.

Fig. 79.

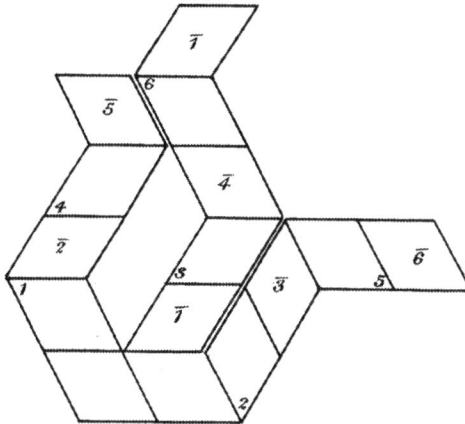

E & F.N.Spon, London & New York

half of the second axis *b b'*, and let *s e*, *s e'*, be the lines of elevation of the angles of the rectangle on the lower half of the same axis.

Lateral plane (*a*).—Take *d e*, Fig. 69, and with one foot of compasses in *d*, Fig. 70, describe the arc *d k*; it shall cut the line *se* in *k*. Upon *d k* describe the rhomboid *k d m n*, which is the lateral plane (*a*) required.

Lateral plane (*β*).—With radius *d d'*, Fig. 69, and foot of compasses in *e'*, Fig. 70, describe the arc *e' f*, cutting the line *s e'* in f. Upon the dotted line *e' f* describe the rhomboid *e' f g h*, which is the lateral plane (*β*) required.

Terminal plane is the base *e d d' e'*, Fig. 69. For the adjustment of planes, see Fig. 73.

RHOMBOHEDRAL SYSTEM.

Axes three, equal in length, intersecting at an angle of 60°; the fourth perpendicular to the other three, and varying in length.

> Forms:—*Rhombohedron, scalenohedron, regular hexagonal prism, regular double six-sided pyramid.*

1. *Rhombohedron.*—A solid, contained under six similar and equal rhombs.* These forms are of two kinds, the *obtuse* and the *acute*, the distinction arising from the structure of the solid angles of the principal axis. In the obtuse variety the solid angles are composed of three faces, the plane angles of which are greater than 90°, Fig. 76; in the acute variety the plane angles of the pyramid are less than 90°, Fig. 77.

The rhombs are adjusted in the same numerical order, and plaited like the square faces of a cube.

Obtuse Rhomobohedron with the plane angles of its faces equal to 60° and 120° an impossible solid.—Let it be required to construct an obtuse rhombohedron, which shall be contained under six faces, the plane angles of which are equal to 60° and 120°. By adjusting these faces according to the key, Fig. 81, a projection like Fig. 79 is obtained. On proceeding to plait this projection into the required form, it is found that by no ingenuity can it be converted into a solid. We have in fact been attempting an impossibility, which results not from any error in the process itself, but from the neglect of a self-evident geometrical truism, that "when three plane angles of 120° meet, they lie in one plane," and hence cannot form a solid angle. What actually does result, is that the figure subsides into a low, flattened pile of

---

* A *rhomb* is that which has its four sides equal, but its angles are not right angles.

rhombs, arranged in the form of a hexagon, Fig. 78. This exceptional rhombohedron was noticed by Brooke.

A second method for projecting the planes of this *quasi* rhombohedron is given in Fig. 79A. It is worthy of notice; for while in its present state it refuses to assume the form of a solid, yet, by simply *creasing its rhombic faces in their short diagonal*, it becomes at once converted into the *regular octahedron*, Fig. 17, which see.

MODE OF USING THE KEY.

Describe a face of any given rhombohedron, Fig. 82, cut it out in cardboard for a pattern, and draw a line across its long diagonal.

Next trace on a sheet of paper successive outlines of the pattern, so that the directions of its diagonals shall correspond with those of the key. When the direction of the *dotted* lines in the key is followed, the model will be an *obtuse rhombohedron;* but when the long diagonal of the pattern corresponds with *continuous* lines on the key, the model is *acute.*

Two distinct forms are thus seen to be produced from one and the same plane by the mere shifting of its sides: it follows that it is not a matter of indifference as to which pair of sides shall meet to form an edge; and, moreover, that, while the process of plaiting determines the relative position of all the faces themselves, it does not locate that of the sides of those faces. Hence the necessity for some simple expedient like that which has been here adopted.

Acute Rhombohedron of Carbonate of Iron.—This solid has two of the faces of an acute pyramid inclined to each other at their edge, at an angle of 67° 20'. (Phillips.)

Plate 43

KEY FOR ADJUSTING THE PLANES OF ALL THE RHOMBOHEDRONS.

Fig. 81.

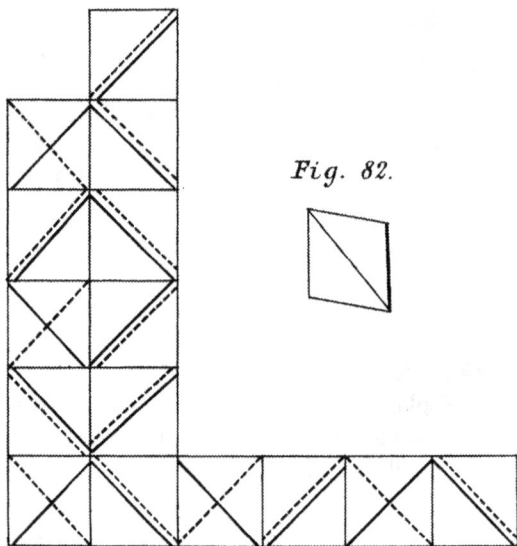

Fig. 82.

Acute ─────────
Obtuse ─────────

Fig. 83.

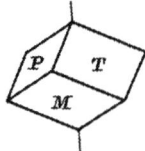

Fig. 84.

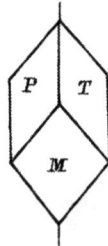

E & F.N.Spon, London & New York

To find the plane angles of a face:—

$$\text{Log. } \frac{1}{2 \sin\frac{1}{2} A} = \cos\frac{1}{2} a.$$

$$\text{Let } A = 67° 20'.$$

$$\frac{1}{2} A = 33° 40'.$$

$$L \sin\frac{1}{2} A = 9.7437921.$$

$$L \ 2\sin\frac{1}{2} A = 10.0448221.$$

$$L \ \frac{1}{2\sin\frac{1}{2} A} = 9.9551779.$$

Hence,

$$\cos\frac{1}{2} A = 25° 36' 85''$$

$\therefore A = 50° 73' 70'' =$ the plane acute angles of the rhomb.

The plane angles of the rhomb, therefore, are equal to

$$50° 73' 70''$$

$$128° 45' 50''$$

To make the figure, the planes are adjusted as in the projection, Fig. 75. See Key.

Obtuse Rhombohedron.

$$\left(\cos\frac{1}{2}A = L\,\frac{1}{2\sin\frac{1}{2}E}\right)$$

Inclination of faces at edges.					Plane angles of faces =			
	°	′		°	′	°	′	°	′
P on T =	92	50	..P on M =	87	10	92	42	87	18
	94	15		85	45	93	57	86	3
	94	46		86	14	94	24	85	36
	105	5		74	55	101	55	78	5
	106	15		73	45	102	38	77	22
	107	20		72	40	103	16	76	44
	106	30		73	30	102	47	77	13
	107	0		73	0	103	5	76	55
	108	30		71	30	103	56	76	4
	126	17		54	43	111	50	68	10
	133	50		46	10	114	9	65	51

Acute Rhombohedron.

$$\left(\cos\frac{1}{2}A = L\,\frac{1}{2\sin\frac{1}{2}E}\right)$$

Inclination of faces at edges.					Plane angles of faces =			
	°	′		°	′	°	′	°	′
P on M =	93	56	..P on T =	86	4	94	14	85	46
	107	30		72	30	115	28	64	32
	109	28		70	32	120	1	60	0

Scalenohedron.—The plane angles of a face of the obtuse rhombohedron of calcite given, to find the scalene triangular face of the scalenohedron.

First, find the *principal axis* of the rhombohedron:—Let Fig. 85 be a rhomb with plane angles equal to 101° 54′ 50″, and let A C be its long diagonal, and B D its short one.

With any side A B, Fig. 85, describe the equilateral triangle E F G, Fig. 86; and let O be its centre.

Produce (Fig. 87) the straight line AX, on which to find the principal axis; and with radius equal to side A B, Fig. 85, describe the arc *c c*, Fig. 87.

Take the line G O, Fig. 86, and set it at right angles at some point along the line A X; it will be found to cut the arc *c c* in B. Then AB, Fig. 87, is equal to a side of the rhomb; B C is equal to its short diagonal; while A C is the *principal* axis.

To make the *scalene triangle* of the scalenohedron:—Produce the line A C, Fig. 87, to S, so that A S is equal to the axis A C. Complete by dotted lines the parallelogram A B C E, and make the *scalene triangle*, Fig. 88:—

      AB = SE, Fig. 87,
      AC = SB, ,,   ,,
      BC = BC, ,, 85.

Double Six-sided Pyramid.—Bounded by twelve equal and similar *isosceles* triangles.

Like the scalenohedron, and belonging to the same system, it is made by joining two of the faces together by their sides, and treating the *trapezium* thus formed as if it were a *rhomb*. Counting two planes as one, they are then adjusted as if for a rhombohedron, numbered in the same order, and made up into a model with the same facility. (Fig. 93A.)

Hexagonal Prism.—Contained under eight faces—six lateral *rectangles*, and two terminal *hexagons*. Inclination of lateral on terminal planes equal to 90°; inclination of lateral planes equal to 120°.

Plate 44

Fig. 85.

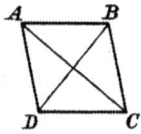

Fig. 86.

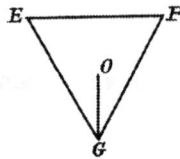

Fig. 87

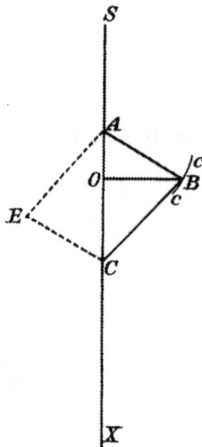

Fig. 88.

E & F.N.Spon, London & New York

Plate 45

SCALENOHEDRON.

Fig. 89.

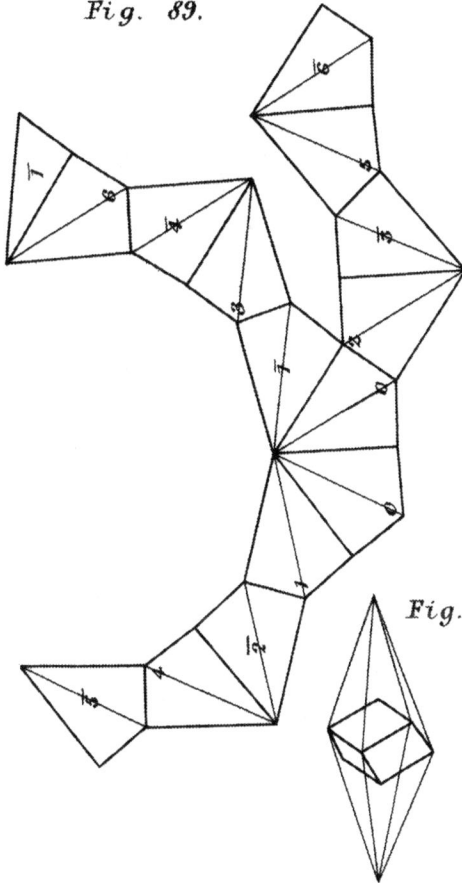

Fig. 89.ª

E & F.N.Spon, London & New York

Plate 46

HALF SCALENOHEDRON AND HALF RHOMBOHEDRON OF CALCITE IN ONE MODEL.

Fig. 90.

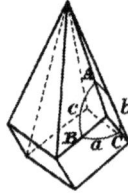

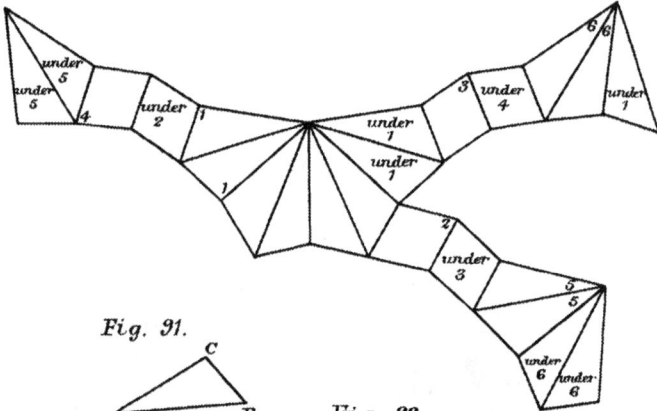

Fig. 91.

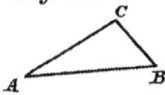

A = 25° 21'
B = 52° 18' 39'
C = 102° 20' 21'

Fig. 92.

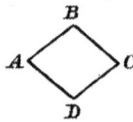

B = 101° 55

E & F.N.Spon, London & New York

Plate 47

DOUBLE SIX-SIDED PYRAMID.

Fig. 93.

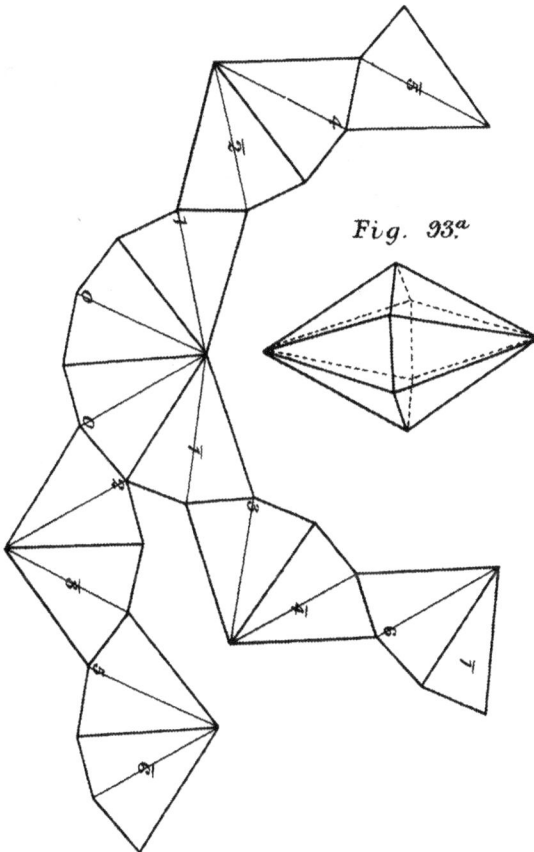

Fig. 93.ª

E & F.N.Spon, London & New York

Plate 48

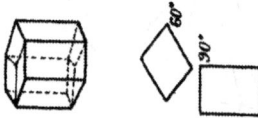

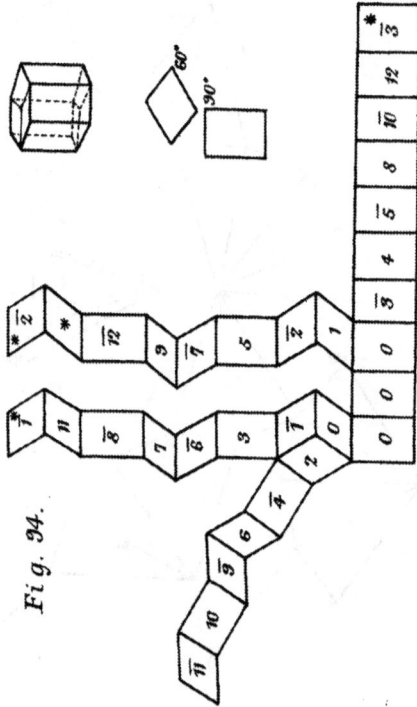

HEXAGONAL PRISM.

Fig. 95.

Fig. 94.

E & F.N.Spon, London & New York

Plate 49

HEXAGONAL PRISM
WITH DOUBLE SIX SIDED PYRAMID.

Fig. 95ª

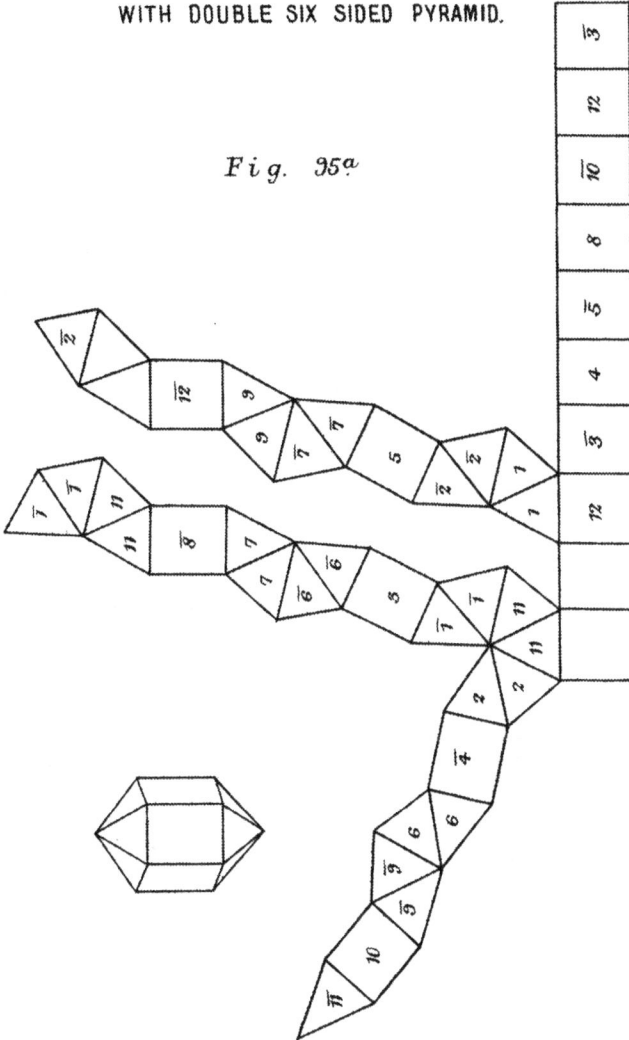

E & F.N.Spon, London & New York

Plate 50

ACUTE   RHOMBOHEDRON.

Fig. 96.

Fig. 98.

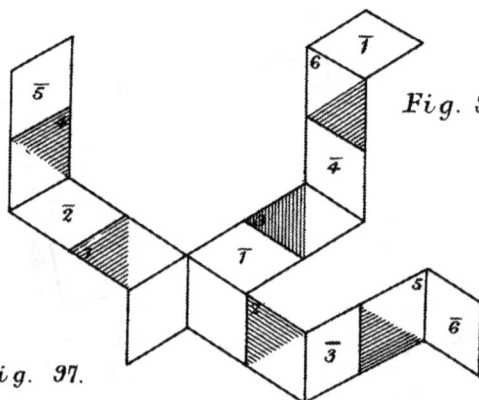

Fig. 97.

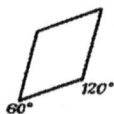

E & F.N.Spon, London & New York

Plate 51

CUBE GROWTH.

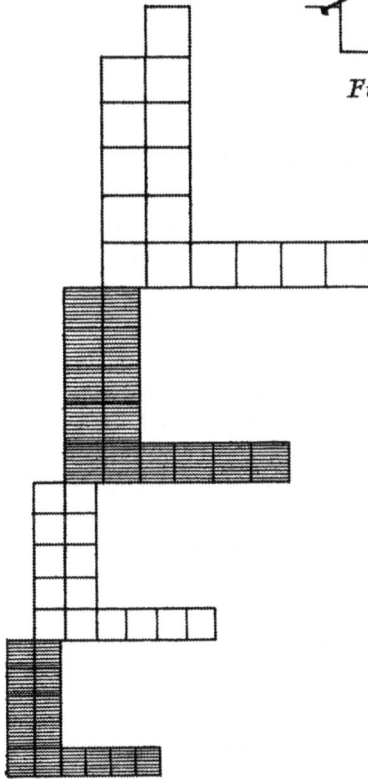

The three cleavages of
Iron pyrites.

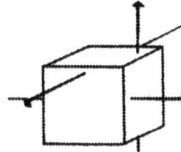

Fig. 99.

Fig. 100.

Instead of using eight planes, it is far more convenient to resolve each terminal hexagon into three rhombs, the plane angles of which are equal to 60° and 120°. The model can now be made by adjusting the planes as in Fig. 94. It will be found to be contained under twelve faces, viz. three rhombs composing each terminal plane, and six rectangles—making twelve in all. By using the type composed of four fillets (Fig. 94), we obtain the model itself in a compact and stable form, but its terminal hexagons present a mosaic of three pieces, all lying in one plane. We may now call a model thus constructed a *typical* hexagonal prism, in as much as by altering the plane angles of the rhombs, pyramids of three and six sides can be raised on its ends, forming modifications which imitate natural crystals, as of quartz, &c. (See Fig. 95A.)

Acute Rhombohedron containing a nucleated *regular octahedron* (Fig. 96):—Let the plane angles of the rhomb, Fig. 97, be equal to 60° and 120°. Make the plane projection, Fig. 98, and shade the faces as in the figure. When the model is completed, the *octahedron* will appear as shown in Fig. 96.

CRYSTAL GROWTH; CLEAVAGE.

Besides mere hollow models, the plaiting process enables us to insert a number of similar forms of *different sizes,* one within the other, so as to resemble real solids composed of concentric layers.

A model made by enclosing cube within cube resembles a real crystal more nearly than one which displays its faces only. It is strikingly suggestive of the mode adopted in the natural growth of a crystal, in its cleavage, and in other structural peculiarities. Crystals are conceived to increase in magnitude by the continual addition of plates of molecules to their surfaces.

The structure of crystals in the order in which their molecules are arranged may be illustrated by an experiment with common salt. If a portion of this salt be dissolved in water, and the water be allowed to evaporate slowly, rectangular crystals will be developed, deposited on the sides and bottom of the vessel. These will at first be very minute, but they will increase in size as the evaporation proceeds; and if the quantity of salt dissolved be sufficient, they will at length attain a considerable magnitude.

If the edge of a knife be applied to the surface of any one of these crystals in a direction parallel to one of its edges, the crystal may, by a slight blow, be cleaved parallel to one of its sides.

It is hence inferred that the molecules of crystals are so arranged as to form plates in the direction of all their primary planes.

This peculiarity in crystal structure is well exemplified in the model, Fig. 100, which is supposed to consist of a series of cubes, small at first, but gradually enlarging from the centre. A model thus constructed will permit of needles being thrust through its entire thickness in directions *parallel to its sides,* but in no other directions.

Parallel markings in straight lines are found on the surfaces of many natural crystals. These are called *striæ* (from the Latin word *stria,* a groove or channel). They are often seen in the cubes of *iron pyrites.* The *striæ* on any given face being always perpendicular to those on its adjacent faces (Fig. 106). This character can be imparted to a model with the greatest ease by drawing parallel lines longitudinally along each fillet (Fig. 105), which will so intersect in plaiting that the model will become an exact fac-simile of the natural crystal (Fig. 106).

These results are suggestive of the primordial arrangement of the molecules in crystal forms. What, for instance, should hinder the particles, when starting into activity, from aggregating themselves into three distinct infinitesimally thin laminæ? destined to grow in three zones; and further, that each zone shall be endued with a force compelling it to bend at a right angle at given intervals, and that these zones shall overlap one another after the manner of a plait, thus forming a minute cubic crystal. The same forces continuing to act, this minute crystal *may* constitute the nucleus around which hundreds—nay, thousands—of fresh laminæ shall develop themselves, each forming a fresh nucleus to the next outer one, until finally the entire form shall consist of an almost innumerable series of hollow cubes, so closely packed as to become virtually a solid.

On this hypothesis the *cleavage* property of a crystal becomes almost a necessary result; the natural joints or intervals of disintegration consisting of the spaces between the nucleated forms of which the crystal is constructed.

Of Crystals of Bismuth and of Bromide and Iodide of Potassium.—It is not usual for crystal forms to disclose to view any part of their structure, excepting that which lies directly on their surface. An opportunity is seldom afforded, therefore, of obtaining a survey of their interior. Hence the initial direction in which the particles are propelled by their respective forces, in order to build up a crystal, is for the most part a mere matter of conjecture. Some remarkable exceptions do occur, however, in certain crystals where nearly the entire face, excepting only a narrow margin near the edge, is absent. These deficiencies are seen to be replaced by deep excavations which frequently descend almost to the centre. Examples of this peculiarity in structure are afforded in the slowly cooled crystals of *bismuth* and in crystals of the *bromide* and *iodide* of *potassium.* The hollows above noticed present the appearance of four-sided pyramids, their apices pointing inwards towards the centre. They appear to consist of a series of square frames piled

Plate 52

RHOMBOHEDRON GROWTH.

Fig. 101.

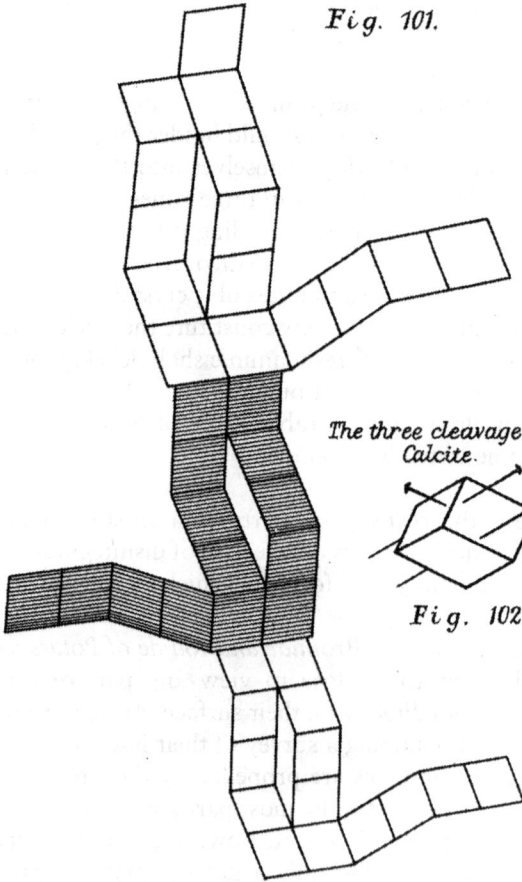

The three cleavages of Calcite.

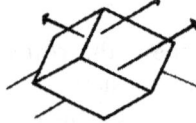

Fig. 102.

E & F.N.Spon, London & New York

Plate 53

OCTAHEDRON GROWTH.

*The 4 cleavages of
Oxydulated Iron or
Native Loadstone.*

Fig. 103.   *Fig. 104.*

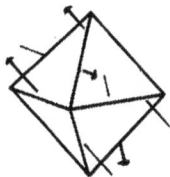

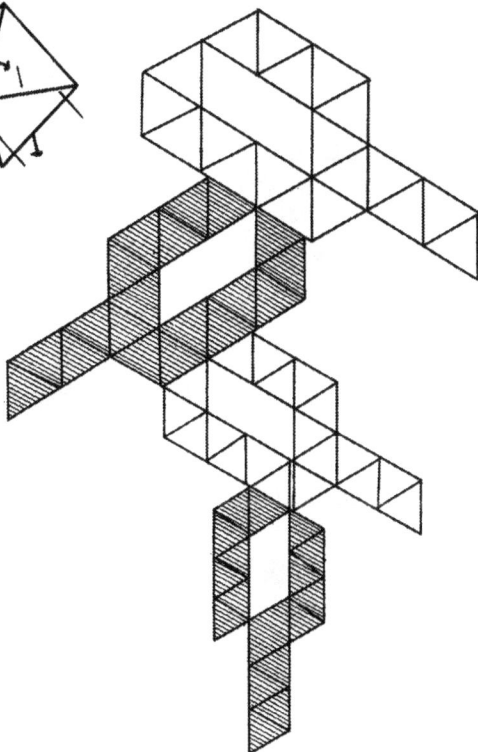

E & F.N.Spon, London & New York

Plate 54

DIRECTIONS *of striation reproduced when longitudinal strice are made on each strip and then plaited into a model.*

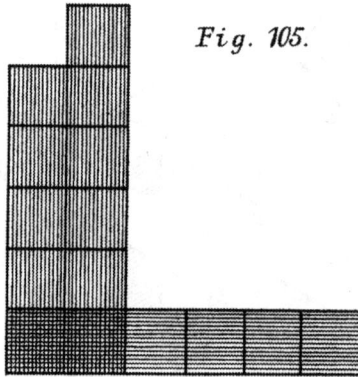

Fig. 105.

CUBE OF IRON PYRITES. *Striæ of adjacent faces perpendicular to each other.*

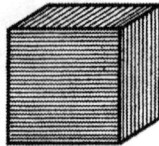

Fig. 106.

E & F.N.Spon, London & New York

Plate 55

CRYSTAL OF BISMUTH.

Fig. 108.

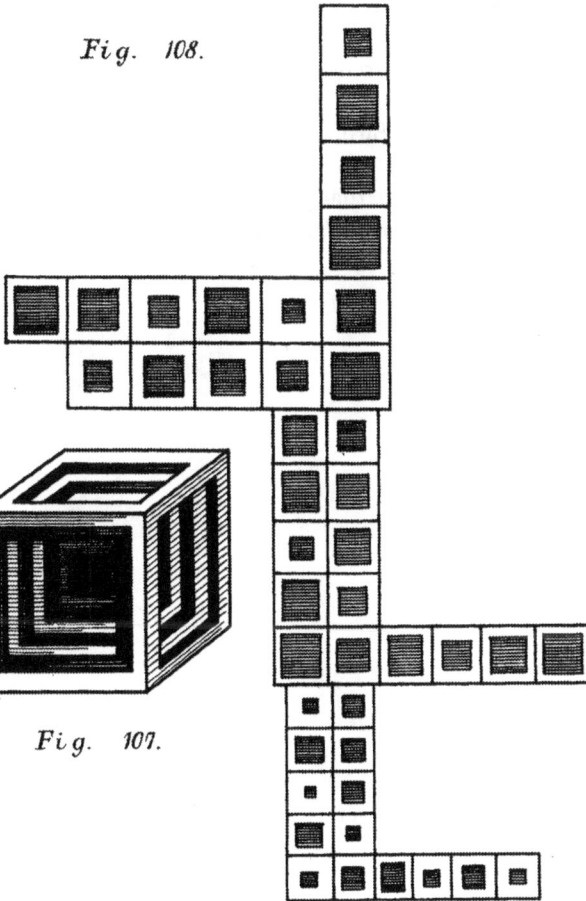

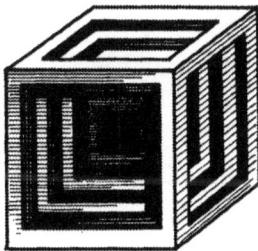

Fig. 107.

E & F.N.Spon, London & New York

up one upon the other, and growing sensibly and abruptly bigger from within outwards; they become graduated, so that each side of the pyramid assumes the appearance of a miniature flight of steps.

From the facilities afforded by the plaiting process for constructing a nucleated model of the cube, it appeared not improbable that by substituting square *frames* for square *faces*, as shown in Fig. 108, models of these beautiful crystals might be made, and if so, that a clue would be thus obtained as to the mysterious process adopted by nature in the formation of structures at once so intricate and incomprehensible.

The resulting model with its projection are shown in the figures.

Even solid crystals, if transparent, may sometimes present certain markings which are suggestive of their development. In some specimens of *uric acid* it was noticed by Dr. Golding Bird, that under the microscope *"they appeared nucleated from the presence of certain internal markings, as if one crystal included another;"* and again, that *" when the rhombic outline of the crystal is replaced by a square one, in such crystals an internal marking like a frame-work is visible."*

www.ingramcontent.com/pod-product-compliance
Lightning Source LLC
Chambersburg PA
CBHW070406200326
41518CB00011B/2091